AF347529

Advanced Technique and Future Perspective for Next Generation Optical Fiber Communications

Advanced Technique and Future Perspective for Next Generation Optical Fiber Communications

Editors

Jian Zhao
Jiangbing Du
Yang Yue
Jianguo Liu

MDPI • Basel • Beijing • Wuhan • Barcelona • Belgrade • Manchester • Tokyo • Cluj • Tianjin

Editors

Jian Zhao
School of Precision
Instruments and
Optoelectronics Engineering
Tianjin University
Tianjin
China

Jiangbing Du
State Key Laboratory of
Advanced Optical
Communication Systems and
Networks
Shanghai Jiao Tong
University
Shanghai
China

Yang Yue
School of Information and
Communications Engineering
Xi'an Jiaotong University
Xi'an
China

Jianguo Liu
Optoelectronics Research and
Development Center
Institute of Semiconductors,
Chinese Academy of Sciences
Beijing
China

Editorial Office
MDPI
St. Alban-Anlage 66
4052 Basel, Switzerland

This is a reprint of articles from the Special Issue published online in the open access journal *Photonics* (ISSN 2304-6732) (available at: www.mdpi.com/journal/photonics/special_issues/NGOFC).

For citation purposes, cite each article independently as indicated on the article page online and as indicated below:

LastName, A.A.; LastName, B.B.; LastName, C.C. Article Title. *Journal Name* **Year**, *Volume Number*, Page Range.

ISBN 978-3-0365-4136-5 (Hbk)
ISBN 978-3-0365-4135-8 (PDF)

Contents

About the Editors

Jian Zhao

Jian Zhao received the B.Eng. degree in Optoelectrical Engineering from Tianjin University, Tianjin, China, in 2003, the M.S. degree in Optics from Nankai University, in 2006, and the Ph.D. degree in Electronics and Information Engineering from The Hong Kong Polytechnic University, in 2010. From 2016 to 2017, he was a visiting scholar of the CREOL, The College of Optics and Photonics, University of Central Florida. He is currently an Associate Professor with the School of Precision Instruments and Optoelectronics Engineering, Tianjin University. He has authored or coauthored more than 50 articles in peer-reviewed journals and leading international conferences. His current research interests include optical fiber communications, artificial intelligence, neural network, optical fiber sensing, and nonlinear optics.

Jiangbing Du

Jiangbing Du received B.S. (02-05) and M.S. (05-08) degrees, respectively, from the College of Physics and Institute of Modern Optics, Nankai University, Tianjin, China. He obtained his PhD degree (08-11) in Electronic Engineering from the Chinese University of Hong Kong. He was with Huawei technologies from 2011 to 2012. He joined Shanghai Jiao Tong University in 2012. Dr. DU is a senior member of both IEEE and OSA. He is the author or coauthor of over 150 journal and conference papers, including 26 OFC/ECOC papers. He has served as TPC/LOC member in a variety of academic conferences, such as OECC, ACP, and so on. His research interests include optical communication, photonic integration, and optical signal processing.

Yang Yue

Yang Yue received B.S. and M.S. degrees in Electrical Engineering and Optics from Nankai University, China, in 2004 and 2007, respectively. He received a Ph.D. degree in Electrical Engineering from the University of Southern California, USA, in 2012. He is a Professor with the School of Information and Communications Engineering, Xi'an Jiaotong University, China. Dr. Yue's current research interests include intelligent photonics, optical communications and networking, optical interconnect, detection, imaging and display technology, integrated photonics, free-space and fiber optics. He has published over 200 peer-reviewed journal papers (including Science) and conference proceedings with >9,000 citations, four edited books, >50 issued or pending patents, >170 invited presentations (including 1 tutorial, >10 plenary and >30 keynote talks). Dr. Yue is a Senior Member of the Institute of Electronic and Electrical Engineers (IEEE). He is an Associate Editor for IEEE Access, and an Editor Board Member for three other scientific journals. He also served as Guest Editor for ten journal special issues, Chair or Committee Member for >80 international conferences, and reviewer for >60 prestigious journals.

Jianguo Liu

Jianguo Liu is currently a Professor at the Institute of Semiconductors, Chinese Academy of Sciences, China. He is the recipient of the National Science Fund for Distinguished Young Scholars, and a national expert on microwave photonic and optoelectronic device integration technology. He has published >130 papers, >100 issued patents, and received one second-class of national technological invention award together with four first-class of provincial/ministerial-level awards.

Editorial

Special Issue on Advanced Technique and Future Perspective for Next Generation Optical Fiber Communications

Jian Zhao [1,*], Jiangbing Du [2], Yang Yue [3] and Jianguo Liu [4]

[1] School of Precision Instruments and Optoelectronics Engineering, Tianjin University, Tianjin 300072, China
[2] State Key Laboratory of Advanced Optical Communication Systems and Networks, Shanghai Jiao Tong University, Shanghai 200240, China; dujiangbing@sjtu.edu.cn
[3] School of Information and Communications Engineering, Xi'an Jiao Tong University, Xi'an 710049, China; yueyang@xjtu.edu.cn
[4] State Key Laboratory of Integrated Optoelectronics, Institute of Semiconductors, Chinese Academy of Sciences, Beijing 100083, China; jgliu@semi.ac.cn
* Correspondence: enzhaojian@tju.edu.cn

Citation: Zhao, J.; Du, J.; Yue, Y.; Liu, J. Special Issue on Advanced Technique and Future Perspective for Next Generation Optical Fiber Communications. *Photonics* **2022**, *9*, 280. https://doi.org/10.3390/photonics9050280

Received: 6 March 2022
Accepted: 13 April 2022
Published: 20 April 2022

Publisher's Note: MDPI stays neutral with regard to jurisdictional claims in published maps and institutional affiliations.

Optical communication refers to the communication mode with optical signal as a carrier. With the development of laser and optical fiber, optical communication has also made great progress and has penetrated into every aspect of our daily life. However, the continuous growth of bandwidth and high-quality service demand has brought unprecedented challenges to optical communication network; thus, it is imperative to find new solutions to meet the new challenges. Due to its powerful computing power, artificial intelligence (AI) is applied in optical communication to improve its performance [1,2]. The combination of artificial intelligence technology and optical communication technology may promise great potential and start a new stage of optical communication and optical network.

This Special Issue aims to discuss advanced analytical tools and new technologies for next-generation fiber optical communication systems and networks. It focuses on the latest advances and future prospects from basic theory to applications, as well as devices, subsystems, and networks. Topics covered include nonlinear fiber optics, advanced devices, AI and deep learning applications in optical fiber communication, and optical network control and management, etc. Seven research articles and one communication paper are included in this Special Issue.

More specifically, passive optical networks are discussed. Zehri, M. et al. proposed a dynamic bandwidth allocation (DBA) algorithm [3]. In time- and wavelength-division multiplexing (TWDM) technology, the laser tuning time (LTT) delay is often ignored when evaluating the performance of dynamic bandwidth allocation (DBA) mechanisms. The DBA algorithm takes LTT into account and is capable of dynamically processing bandwidth and converting the laser wavelength of an optical network unit. By introducing the longest processing time, the first scheduling discipline (LPT) algorithm can reduce the queue delay by 73% compared to interleaved polling with an adaptive cycle time (IPACT) and 33% compared to the WFQ algorithm, which has high practical significance.

Free-space optical communication (FSOC) has the advantages of fast communication speed, strong anti-interference ability, and high security. The use of a FSOC system on unmanned aerial vehicle (UAVs) will help improve communication capabilities and eliminate its strict requirements for electromagnetic environment. Zhang, Y. et al. established an electromagnetic immune FSOC system for UAV control link [4]. The system has great anti-turbulence, anti-vibration, and anti-flight interference performance. Moreover, it achieves the goal of miniaturization without using a gimbal mirror system, a beacon camera system, and a four-quadrant photodetector (QPD), allowing the system to be used in situations where small FSOC devices are required, such as satellite–submarine communication.

In the field of improving optical communication performance, Yin, H. et al. proposed a method to realize frame synchronization based on Hough transform (HT) methods [5].

Hough transform, a classical method for line detection in digital image processing, is applied to fringe detection in intercept matrix. Simulation results in a 56 Gbps QPSK coherent optical transmission system show that the proposed algorithm can achieve frame synchronization, even with high bit error rate of 0.3. In order to improve the transmission power efficiency, Sampath, K.I.A. et al. validated three previously proposed methods for reducing the peak-to-average power ratio (PAPR) of OSSB-SC signals: peak folding, peak clipping, and high-pass Hilbert transform methods [6]. In this study, he numerically compared the results of the three methods in a 10 Gbit/s non-return-to-zero (NRZ)-coded 100 km-single channel transmission link. The self-phase modulation (SPM) threshold induced by peak folding and peak clipping increased by 2.40 dB and 2.63 dB due to the decrease in the peak-to-average ratio, while the SPM threshold induced by high-pass Hilbert transform increased by 9.86 dB. Moreover, peak folding can suppress the driving signal noise. These methods bring a new development direction to improve the transmission power efficiency of OSSB-SC. Additionally, Hayal, M.R. et al. proposed a model of passive optical network (PON) wavelength division multiplexing (WDM) technology based on hybrid fiber FSO (HFFSO) link, which has shown great transmission performance under the condition of 20 gbit/s-4000 m [7]. In this design, M-ary DPPM-M-PAPM modulation greatly improves performance since it can offer extra information bits. They also studied the influence caused by the turbulence effect on the proposed system based on OOK-M-ary PAPM-DMPP modulation. Simulation results confirm that OOK/M-ary DPPM-M-PAPM hybrid optical modulation technique can be applied in the DWDM-FSO hybrid links for optical wireless and fiber-optic communication systems to improve efficiency significantly, as well as reduce the effect of AT, ICC, and ASE noise.

When discussing the combination of artificial intelligence technology and optical communication, Chang, S. H. took the technical network of machine learning in optical communication as the research object to explore the key technologies and application development of machine learning in optical communication [8]. Through patent analysis, he built a network model that can predict technology development trends, which provided a reference for industry development.

Li, C. et al. proposed and theoretically proved the internal dynamics of complex dissipative soliton-bound states using deep convolution networks (DCNs) [9]. They demonstrated the results via VGG, ResNets, and DenseNets, and achieved high-precision results. They extracted phase evolution information of more complex soliton molecules from time-stretch-dispersive Fourier transform (TS-DFT) spectral data by modifying the network structure. When extracting the dynamics information of complex five-soliton molecules from TS-DFT data, the 48-layer DenseNet obtained a mean Pearson correlation coefficient (MPCC) of 0.9975, which showed the best performance. ResNet and VGG have MPCCs of 0.9906 and 0.9739, respectively. It can be seen from the results that these methods are universal in extracting the internal information of complex soliton molecules with high accuracy. Additionally, Zhang, L. et al. proposed a new architecture combining time lenses and optical neural networks (TLs-ONNs) based on the optical space–time duality [10]. Time lenses were applied to the neural network to achieve imaging of time signals by controlling the phase information of time lens. The performance of the network is tested by simulation, and the accuracy of speech recognition is 95.35%. By fusing the ONN with the photon time stretching test system, not only can real-time data processing be achieved, but also the power and cost consumption can be reduced. This architecture is expected to make a breakthrough in initial screening of cancer cells and can find widespread application in high-throughput data processing.

The continuous development of cloud computing, Big Data, Internet of Things, and other emerging businesses has brought about the explosive growth of data volume, which has created new development challenges in optical fiber communication. In order to meet the demand of a large amount of data processing, the existing communication system needs to improve the stability and transmission efficiency. In addition, emerging technologies, such as artificial intelligence (AI), can also provide a new development direction for optical fiber optic communication. Artificial intelligence technology in optical communication and optical network is still in its infancy, but has made some progress in deep convolution networks (DCN) and time lens–optical neural networks (TL-ONNs). Whether we continue to improve the performance of traditional optical fiber communication systems or combine emerging technologies to find new bases, the next generation of optical fiber communication system will develop towards the direction of high speed, high stability, high transmission efficiency, and intelligence.

Conflicts of Interest: The authors declare no conflict of interest.

References

1. Wu, X.; Hu, F.; Zou, P.; Lu, X.; Chi, N. The performance improvement of visible light communication systems under strong nonlinearities based on Gaussian mixture mode. *Microw. Opt. Technol. Lett.* **2020**, *62*, 547–554. [CrossRef]
2. Chi, N.; Zhao, Y.; Shi, M.; Zou, P.; Lu, X. Gaussian kernel-aided deep neural network equalizer utilized in underwater PAM8 visible light communication system. *Opt. Express* **2018**, *26*, 26700–26712. [CrossRef] [PubMed]
3. Zehri, M.; Haastrup, A.; Rincón, D.; Piney, J.R.; Sallent, S.; Bazzi, A. A QoS-Aware Dynamic Bandwidth Allocation Algorithm for Passive Optical Networks with Non-Zero Laser Tuning Time. *Photonics* **2021**, *8*, 159. [CrossRef]
4. Zhang, Y.; Wang, Y.; Deng, Y.; Du, A.; Liu, J. Design of a Free Space Optical Communication System for an Unmanned Aerial Vehicle Command and Control Link. *Photonics* **2021**, *8*, 163. [CrossRef]
5. Yin, H.; Li, S.; Huang, Z.; Chen, J. A Novel Data-Aided Frame Synchronization Method Based on Hough Transform for Optical Communications. *Photonics* **2020**, *7*, 65. [CrossRef]
6. Sampath, K.I.A.; Takano, K.; Maeda, J. Peak-to-Average Power Ratio Reduction of Carrier-Suppressed Optical SSB Modulation: Performance Comparison of Three Methods. *Photonics* **2021**, *8*, 67. [CrossRef]
7. Hayal, M.R.; Yousif, B.B.; Azim, M.A. Performance Enhancement of DWDM-FSO Optical Fiber Communication Systems Based on Hybrid Modulation Techniques under Atmospheric Turbulence Channel. *Photonics* **2021**, *8*, 464. [CrossRef]
8. Chang, S.-H. Patent Technology Network Analysis of Machine-Learning Technologies and Applications in Optical Communications. *Photonics* **2020**, *7*, 131. [CrossRef]
9. Li, C.; He, J.; Liu, Y.; Yue, Y.; Zhang, L.; Zhu, L.; Zhou, M.; Liu, C.; Zhu, K.; Wang, Z. Comparing Performance of Deep Convolution Networks in Reconstructing Soliton Molecules Dynamics from Real-Time Spectral Interference. *Photonics* **2021**, *8*, 51. [CrossRef]
10. Zhang, L.; Li, C.; He, J.; Liu, Y.; Zhao, J.; Guo, H.; Zhu, L.; Zhou, M.; Zhu, K.; Liu, C.; et al. Optical Machine Learning Using Time-Lens Deep Neural NetWorks. *Photonics* **2021**, *8*, 78. [CrossRef]

Article

A QoS-Aware Dynamic Bandwidth Allocation Algorithm for Passive Optical Networks with Non-Zero Laser Tuning Time

Mohammad Zehri [1,2], Adebanjo Haastrup [1], David Rincón [1,*], José Ramón Piney [1], Sebastià Sallent [1] and Ali Bazzi [2]

[1] Department of Network Engineering, Universitat Politècnica de Catalunya (UPC), 08860 Castelldefels, Spain; mohammad.habib.zehri@upc.edu or mohamad.zehri@liu.edu.lb (M.Z.); adebanjo.haastrup@upc.edu (A.H.); jose.ramon.piney@upc.edu (J.R.P.); sallent@entel.upc.edu (S.S.)

[2] Department of Computer and Communication Engineering, Lebanese International University (LIU), Beirut 14404, Lebanon; ali.bazzi@liu.edu.lb

* Correspondence: david.rincon@upc.edu; Tel.: +34-93-413-7056

Abstract: The deployment of new 5G services and future demands for 6G make it necessary to increase the performance of access networks. This challenge has prompted the development of new standardization proposals for Passive Optical access Networks (PONs) that offer greater bandwidth, greater reach and a higher rate of aggregation of users per fiber, being Time- and Wavelength-Division Multiplexing (TWDM) a promising technological solution for increasing the capacity by up to 40 Gbps by using several wavelengths. This solution introduces tunable transceivers into the Optical Network Units (ONUs) for switching from one wavelength to the other, thus addressing the ever-increasing bandwidth demands in residential broadband and mobile fronthaul networks based on Fiber to the Home (FTTH) technology. This adds complexity and sources of inefficiency, such as the laser tuning time (LTT) delay, which is often ignored when evaluating the performance of Dynamic Bandwidth Allocation (DBA) mechanisms. We present a novel DBA algorithm that dynamically handles the allocation of bandwidth and switches the ONUs' lasers from one wavelength to the other while taking LTT into consideration. To optimize the packet delay, we introduce a scheduling mechanism that follows the Longest Processing Time first (LPT) scheduling discipline, which is implemented over the Interleaved Polling with Adaptive Cycle Time (IPACT) DBA. We also provide quality of service (QoS) differentiation by introducing the Max-Min Weighted Fair Share Queuing principle (WFQ) into the algorithm. The performance of our algorithm is evaluated through simulations against the original IPACT algorithm, which we have extended to support multi-wavelengths. With the introduction of LPT, we obtain an improved performance of up to 73% reduction in queue delay over IPACT while achieving QoS differentiation with WFQ.

Keywords: passive optical networks; laser tuning time; TWDM; Dynamic Bandwidth Allocation; optical access networks; scheduling

---

Citation: Zehri, M.; Haastrup, A.; Rincón, D.; Piney, J.R.; Sallent, S.; Bazzi, A. A QoS-Aware Dynamic Bandwidth Allocation Algorithm for Passive Optical Networks with Non-Zero Laser Tuning Time. *Photonics* **2021**, *8*, 159. https://doi.org/10.3390/photonics8050159

Received: 30 March 2021
Accepted: 30 April 2021
Published: 10 May 2021

Publisher's Note: MDPI stays neutral with regard to jurisdictional claims in published maps and institutional affiliations.

1. Introduction

Data traffic is increasing massively as the type of services is changing interactively towards media and streaming. It has been predicted that connected devices will grow significantly to reach 100 billion devices by year 2025 [1]. Passive Optical Networks (PON) are one of the major driving forces behind residential broadband access and 5G networks [2] (for example, in cloud radio access networks (C-RAN) [3,4]), as they meet the ever-increasing demand for bandwidth-intensive applications such as ultra-high-definition TV, immersive video, and the stringent end-to-end latency required by mission-critical applications.

PON is a cost-effective optical technology that provides the advantage of using passive network elements to connect users in access networks. They consist of a central unit called an Optical Line Terminal (OLT) at the central office of the internet service provider, and it connects through optical fiber to several Optical Network Units (ONUs) located in—or close to—the customers' premises within a 20 km range [5]. The optical fiber cable is shared by introducing passive optical splitters into the optical distribution network (ODN) located between the OLT and the ONUs, by which it reaches up to 64 users [6] (although some architectures allow split ratios of up to 1024 [7]). In the upstream direction, the splitter combines the upstream wavelengths from the ONUs to the OLT [8]. The PON architecture is referred to as a point-to-multipoint (P2MP) system [9]. It is a very cost-effective and easy-to-manage solution, as it does not require any active electronic devices between the OLT and the ONUs [10].

The PON system is based on a shared model that allows bi-directional communication between the ONUs and the OLT. The downstream traffic is broadcast from the OLT to all ONUs while the upstream communication from the ONUs to the OLT is achieved using a time-sharing principle [8]. Owing to the shared nature of the PON, and ONU systems competing for network capacity, a mechanism must be put in place to control the allocation of the upstream transmission capacity in real time, thus avoiding data collision if two or more ONUs transmit simultaneously towards the OLT. PONs employ a Dynamic Bandwidth Allocation (DBA) algorithm to orchestrate the allocation of network resources in the shared medium. One of the main requirements of a DBA is that it satisfies the low latency and huge bandwidth requirements of emerging applications [11].

The new generation of PON technology is based on the Time- and Wavelength-Division Multiplexing (TWDM) technique, which has been described as an evolutionary step that allows using multiple wavelengths to increase the capacity of the PONs [12]. TWDM is a hybrid technique that combines Wavelength-division multiplexing (WDM) capacity expansion with the inherent resource granularity of a Time-division multiplexing (TDM-PON) to meet the growing demands for bandwidth, reach and aggregation [13]. Some TWDM-PON proposals, based on four wavelengths typically have a maximum throughput capacity of 40 Gbps, thus meeting the requirements of the NG-PON2 standards [4,14]. TWDM-PON is used as a major application in mobile fronthaul networks for connecting the centralized baseband unit (BBU) and remote radio heads (RRHs) in 5G C-RANs, which have extreme requirements in terms of capacity, latency, and cost-efficiency [15,16].

In TWDM-based PONs, the resource allocation process in the upstream link is two-dimensional, consisting of wavelength and time slot allocation. The DBA scheme dynamically allocates the wavelengths (typically four) among the ONUs and shares available bandwidth in terms of time slots among the ONUs in the upstream link. An important characteristic of TWDM-PONs is the use of tunable transceivers at the ONUs [5], which are thus enabled to switch their wavelengths. It is important for the DBA to efficiently handle the assignment of the wavelengths, which involves the switching of ONUs from one wavelength to the other. The wavelength assignment decision is communicated to the ONUs by the OLT, and the ONUs can transmit their frames at their allotted time slots on the assigned wavelength [5]. This approach makes it necessary for ONUs to change their wavelengths to optimize the use of the shared medium. ONUs use tunable lasers to facilitate the switching of the wavelengths as instructed by the OLT, thus adding both complexity and a Laser Tuning Time (LTT) delay that may have a great impact on the performance of the system [4,5]. Only a few research works consider LTT when designing or evaluating the performance of DBAs for TWDM networks [17,18]. It is therefore, necessary to develop more sophisticated DBA algorithms that will ensure fair distribution of resources among ONUs while taking into consideration the delays caused by lasers switching between wavelengths.

In this paper, we propose a novel DBA algorithm to efficiently manage the allocation of bandwidth and wavelength assignment while considering the LTT delay. Transmitting on multi-wavelength PON poses the problem of scheduling with a constraint on the completion time and an overall effect on the delay. Therefore, we aim to reduce the queue delay by introducing a scheduling scheme based on the Longest Processing Time first (LPT) principle [19]. The goal of LPT is to minimize the maximum completion time for processing and transmitting the requests from the ONUs. This is achieved by the OLT sorting the ONUs' bandwidth requests in descending order, with the largest request being processed first. Finally, we introduce weight-based QoS differentiation following the Max-Min Weighted Fair Share principle [20] to ensure a guaranteed bandwidth for demands requested by the users since traditional IPACT algorithm does not guarantee QoS [21].

The main contribution of our TWDM-DBA is to effectively reduce end-to-end delay, and efficiently utilize the bandwidth while achieving QoS differentiation. We validated our algorithm by comparing it with the traditional IPACT algorithm, which has been extended to use up to four wavelengths. The performance metrics of our study include queue delay and throughput. The results show that our proposed DBA can significantly improve network performance in terms of queue delay and throughput while adding QoS differentiation.

The remainder of the paper is organized as follows. Related work and the state of the art are summarized in Section 2. Section 3 introduces the proposed TWDM-DBA algorithm. Section 4 describes a performance evaluation of the proposed approach using simulation results. Conclusions and future work are described in the last section.

2. Related Work

2.1. PON Standards

The International Telecommunication Union (ITU) and the Institute of Electrical & Electronics Engineers (IEEE) are active players that have been at the forefront of developing PON standards over the past twenty years [22].

IEEE introduces the concept of Ethernet over the shared media of Passive Optical Networks, called Ethernet PON (EPON) standards [8]. These standards fall within the IEEE 802.3 standards series [23], and they have continued to evolve. The initial gigabit speeds of EPON provides 1 Gb/s offering a symmetric service to 64 customers with one strand of fiber from the central office (CO) to end users over a maximum distance of 20 km [24]. Later, 10G-EPON standardized the symmetric service at 10 Gb/s [25]. One of the latest versions of the standards is known as NG-EPON under the IEEE 802.3ca standards and aims to support more capacity, 25 Gb/s and 50 Gb/s on 25 Gb/s serial streams and improve data transmission efficiency in the access network by using multiple wavelengths [26].

The ITU developed its variant of PON known as Gigabit-capable Passive Optical Networks (GPON), which handles Asynchronous Transfer Mode (ATM) packets and GEM (GPON Encapsulation Method) frames while providing QoS assurance and 2.5/1.25 Gb/s asymmetric service [22]. The latest version of ITU GPON standards is ITU-T G.989.2, which specifies Next-Generation Passive Optical Network 2 (NG-PON2) [27]. It has three types of channel rates in each of the wavelengths: 10/2.5 Gbps, 10/10 Gbps, and 2.5/2.5 Gbps—downstream and upstream. NG-PON2 introduces time and wavelength division multiplexing (TWDM), which aggregates multiple wavelengths to achieve increased capacity with a nominal aggregate capacity of 40 Gbps in the downstream direction and 10 Gbps in the upstream direction [27,28]. The G.HSP project under the ITU-T Study Group 15 is currently working on providing 50 Gb/s on a serial stream.

For a long time the efforts of the standardization groups have focused on increasing the rate (from 1 to 50 Gbps), maintaining the same aggregation rate (16/32/64 customers) and the same reach between the central office (CO) and the customers. To avoid the reach limitation, the ITU-T proposes to use mid-span reach extenders that involve deploying remote active cabinets resulting in an additional deployment and maintenance cost. This solution has been standardized for G-PON [29] and XG(S)-PON [30]. The limitation of the aggregation rate is not limited by the upper layers of protocols, which in XG(S)-PON can manage up to 1024 ONU identifiers, but by the limitations of the physical layer, and specifically the power budget.

To overcome the current reach and aggregation limitations of PON networks, a working group was formed in November 2018 under the auspices of IEEE, the P802.3cs "Super-PON" [25]. Subsequently, the ITU-T initiated the ITU-T Q2/SG15 working group and approved the standardization of Super-PON in 2019 that will be defined in ITU-T recommendation G.9807.3 [31]. Under the Super-PON umbrella, it is proposed to create a PON network with a range of up to 50 km, with a coverage of 1024 customers per fiber over a passive optical distribution network (ODN). In the optical layer, the Super-PON will carry multiple optical carriers over one fiber strand which will allow multiple PON instances to be mapped to different carriers. This is possible by combining WDM and TDMA technology, optical amplification and multiplexing performed in the Central Office (CO), wavelength routing in the ODN and the fully tunable transmitters in the ONU [32].

2.2. Laser Tuning Time

Tunable optical components in the ONUs enable extensive wavelength flexibility and allow the ONUs to change from heavily loaded wavelengths to idle ones in order to reduce delay and create resource allocation balance in the system [11]. There are several types of tunable lasers, which are categorized based on the speed at which they can switch wavelengths [4,5]. Three classes are defined by ITU-T [27], as shown in Table 1. The lasers can switch between wavelengths within a few microseconds and one second. While the slow lasers are considerably cheaper, the fast lasers are very costly and energy-consuming [12]. Some solutions for a fully tunable ONU transmitter are distributed Bragg reflector (DBR) lasers and Distributed feedback (DFB) lasers. This second solution is more developed and has a lower cost, although due to problems in temperature stabilization, in WDM systems, it is common to limit tuning to four channels [32].

Table 1. Classes of Laser Tuning Time, according to ITU-T [23].

Class	Laser Tuning Time
Class 1	<10 µs
Class 2	10 µs to 25 ms
Class 3	25 ms to 1 s

2.3. DBAs for TWDM-PONs

The issue of the OLT allocating bandwidth to the ONUs is resolved by following a layered approach for job scheduling, as proposed in [33]. The scheduling concept can be divided into scheduling framework and scheduling policy. The scheduling framework deals with the OLT making scheduling decisions for job processing, and the scheduling policy addresses how the DBA allocates time and wavelengths to the ONUs. There are three ways of implementing the scheduling framework: Online, Offline, and Just-in-Time. Online scheduling allows bandwidth to be allocated to ONUs' jobs as soon as they are received at the OLT. Since requests are granted to the ONUs immediately without waiting to consider requests from other ONUs, the system is considered unfair. In offline scheduling, decisions are made after all the requests from the ONUs are received at the OLT, thereby assuring fairness in the system. The downside of the offline scheme is the delay and the underutilization of the link during the time when the OLT receives the report from the first ONU and the time when it issues the grant. To address the aforementioned problems, Just-in-Time scheduling was proposed by [33] to allow the OLT to postpone decision-making until one channel is about to become idle. The decision-making in Just-in-Time scheduling occurs later than in online scheduling and sooner than in offline scheduling.

Many studies have been carried out on the Dynamic Bandwidth Allocation (DBA) algorithms for TWDM-PONs. The DBA algorithms in TWDM-PON are two-dimensional, dynamically combining bandwidth and wavelength allocation for the ONUs in the PON system. This DBA scheme is referred to as Dynamic Wavelength and Bandwidth DWBA algorithms [17]. DWBA allows dynamic wavelength and dynamic time slots allocation to the ONUs by maintaining information about each wavelength channel and determines the transmission period assigned to each ONU on a specified wavelength channel [8]. DWBA can be implemented by using either separate time and wavelength scheduling (STWS) algorithms or joint time and wavelength scheduling (JTWS) algorithms [12]. STWS algorithms decouple the wavelength assignment from the time slot allocation and are thus simple. JTWS algorithms, on the other hand, combine the wavelength assignment with time slot allocation. JTWS is more efficient and scalable but more complex than STWS [34,35]. The work of [36] decouples the wavelength assignment from the Time-Division Multiple Access (TDMA) assignment and removes scheduling complexity by using an adaptive threshold, thus leading to a lower processing requirement. An adaptive wavelength allocation pattern for scheduling multi-wavelength ONUs in NG-EPON is proposed in [37] to achieve a low packet-loss ratio. The wavelength allocation pattern is based on an adaptive threshold that reflects both the ONU's absolute bandwidth request size and the relative bandwidth request size sent to other ONUs.

Some of the recent works on DBA concern specific network architectures. The work of [4] is on mobile fronthaul, with attention given to the bursty nature of its traffic, and proposes a DBA able to satisfy the strict latency requirements. A Dynamic Wavelength and Bandwidth Allocation (DWBA) scheme for TWDM-PON is proposed in [11] to satisfy the strict delay requirement for fronthaul with a minimum number of active wavelength channels. It minimizes active wavelength channels by considering the high burstiness of fronthaul data transmission and using the difference in the propagation delay between the OLT and ONUs. A low latency DBA scheme for NG-PON2 to support both 5G fronthaul services and data services is proposed in [38]. The DBA splits the upstream frame into sub-frames of equal duration, and each ONU is allocated with a time slot in each sub-frame without specifying an allocation interval for the associated queue.

The DBA in TWDM-PON requires wavelength tuning functions; however, most of the dynamic wavelength assignment algorithms do not consider the tuning time. Wavelength tuning reduces channel utilization and increases packet delay [18]. The work of [17] is, to the best of our knowledge, the first to consider the laser tuning time in the design of the DBA. It notes that proper DBA algorithms should maximally exploit the statistical gain among requests, under the condition that lasers are given enough time to switch wavelengths. A DBA based on a minimum wavelength tuning (MWT) scheme is proposed in [39]. The scheme minimizes the frequency of wavelength tuning to reduce packet delay and improve channel utilization. The DBWA proposed in [18] manages transceivers with tunable lasers that have different LTT values combined with transceivers that have non-tunable (fixed) lasers in a single transmission. A multi-tuning-time ONU scheduling (MOS) algorithm that uses an adaptive scheduling algorithm for the coexistence of ONUs with different tuning time in virtual PON is proposed in [40]. The MOS algorithm is able to reduce waste of bandwidth resources and achieve load balancing.

Concerning the quality of service (QoS) in TWDM-PON, some works have recently started to appear in the literature, but they are very limited. A high-priority first dynamic wavelength and bandwidth allocation algorithm in TWDM-PON is proposed in [41]. The DWBA is implemented using five types of transmission containers (T-CONTs), and it can execute four kinds of bandwidth strategies to effectively reduce the average delay, slightly improve bandwidth utilization, and ensure greater fairness for the diverse types of traffic. A QoS-based DWBA for multi-scheduling domain EPON (MSD-EPON) is proposed in [8] to arbitrate the channel bandwidth efficiently in NG-EPON. Following an innovative approach according to the modern traffic requirements, it uses an adequate technique by deploying a blend of online-offline according to the traffic types. A max-min fair allocation scheme is introduced into the algorithm proposed in [42] to provide a minimum level of service in every frame. This technique uses a combination of status reporting and traffic monitoring techniques to achieve fairness and a significant decrease in the average delay.

Existing DBA algorithms as discussed above, address different individual and specific problems such as delay, bandwidth utilization and QoS etc. in PONs without any of them addressing all the issues collectively. There is therefore, a need for a new DBA algorithm to be developed taking LTT and QoS into consideration while minimizing the delay and efficiently utilizing the bandwidth. Our work is based on IEEE EPON standards while drawing on several ideas, such as multiwavelength capability and the associated LTT delay introduced in ITU standards under NG-PON2, which allows ONUs to change their wavelengths. Our contributions align with the work of [17,39] on the application of LTT in DBA algorithms, and we go further by focusing on maintaining a balance between the switching of the ONUs' wavelengths and the associated delay, thus allowing us to achieve optimal bandwidth utilization. We reduce the frame makespan and minimize the delay by introducing the LPT scheme, which is a member of the bin-packing method that is similar to the MULTIFIT used on IPACT in [17] for scheduling the requests from the ONUs.

Additionally, we apply a weight-based bandwidth guarantee scheme in accordance with the Max-Min Weighted Fair Share principle in order to assure QoS differentiation in multi-wavelength PONs like NG-PON2. The Max-Min Weighted Fair Share principle is based on maximizing the minimum share of the ONU whose demand has not been satisfied. Our proposal extends the basic max-min fair allocation proposed in [42] by using priority based on weight in order to capture users' service requirements and weights accordingly. To the best of our knowledge, this is the first time an LTT-aware DBA is enhanced with LPT and WFQ.

3. The Proposed Algorithm

Our proposal builds on IPACT DBA [21], an online algorithm that follows an interleaved polling scheme to schedule transmission from the ONUs in a centralized approach. The requests from the ONUs are sent to the OLT, which has complete knowledge of the queue of the ONUs and when the last bit will arrive. With this knowledge, the OLT will start scheduling the grant for the next ONU. Since the OLT does not have to wait for the rest of the ONUs' requests to reach the OLT before it starts processing them, the waiting time is reduced and the overall delay is minimized.

The original IPACT algorithm has been extended with the capability of coping with multiple wavelengths of the TWDM-PON in [43]. Optimally scheduling the requests from the ONUs on the four wavelengths in TWDM-PON is a problem similar to the scheduling of computational tasks in a multiprocessor environment with identical processors acting in parallel. Mapping this environment with multiprocessor scheduling, with wavelength channels as machines and ONUs' requests as jobs, is indicative of an NP-hard optimization problem, which is computationally prohibitive [44]. Given a set J of jobs where job J_i has length L_i and several wavelengths ω, our objective is to achieve the earliest possible time required to schedule all jobs in J on ω wavelengths such that none overlaps. Since there is a large number of requests coming from the ONUs to be transmitted on the four wavelengths in real-time, heuristic approaches are most suitable in achieving near-optimal scheduling efficiency [19].

We introduce the LPT scheduling algorithm, due to its simplicity, to solve the problem of scheduling the requests on multiple wavelengths to achieve minimal makespan of the requests' processing [45,46]. LPT is a non-preemptive scheduling algorithm that uses the priority to schedule requests to achieve near-optimal efficiency. LPT allows the sorting of the requests made to the OLT during a cycle i by the ONUs $J_1(i), J_2(i) \ldots J_M(i)$, according to the length of time needed for them to be processed such that $L_r(i) \geq L_s(i) \geq \ldots \geq L_m(i)$ being r, s, and $m \leq M$. LPT has the advantage of scheduling almost equal loads on the wavelengths and avoiding situations where some wavelengths will be idle. The upper limit of LPT, $\frac{C_{max}(LPT)}{C_{max}(OPT)}$, has the approximation ratio shown in (1) where $C_{max}(LPT)$ is the maximum makespan of LPT heuristic, and $C_{max}(OPT)$ is the maximum makespan of an optimal scheduler [47].

$$\frac{C_{max}\ (LPT)}{C_{max}(OPT)} \leq \frac{4}{3} - \frac{1}{3M} \tag{1}$$

At the beginning of each cycle, the algorithm acknowledges the number of connected ONUs whose queues are not empty. Based on the lengths of the jobs, the jobs reported from connected ONUs (J_m) are sorted in descending order. The ONUs are then assigned to the respective available wavelengths ω such that ONU m with job J_m *(i)* with the longest processing time $L_m(i)$ is processed first and followed by the next one, assigned to the minimally loaded channel. If $\sum_0^M L_m \leq \delta_{max}$, the requested time is granted for the connected ONUs in a cycle (i), else δ_{max} will be granted and certain jobs with lower lengths have to wait for the cycle ($i+1$). The aforementioned parameters are summarized in Table 2, and the pseudocode is provided in Algorithm 1.

Table 2. Parameters of the LTT-aware QoS based algorithm.

Parameter	Description
M	Total number of ONUs
Ω	Assigned wavelength, $0 \leq \omega \leq 3$
i	Cycle number i, $0 < i < \infty$
$J_m(i)$	The job requested by ONU m at cycle i, $1 < m < M$
$L_m(i)$	Length of job $J_m(i)$ requested by ONU m at cycle i
T	Laser tuning time
ω_i	The wavelength assigned for ONU m during cycle i
$\theta(\omega, i)$	Waiting time for a job $J_m(i)$ on a wavelength ω during cycle i
δ_{max}	Maximum allowed cycle length in bytes
φ_ω	The completion time of the last job on wavelength ω
α_m	Weight of the ONU m
α_{min}	The smallest weight among ONUs
Γ	The summation of the normalized weights
P	Weighted base resource share for an ONU
k	Number of connected ONUs in cycle i
β_m	Resource share based on the weight of each user

Furthermore, to guarantee fairness in the sharing of resources as IPACT has no inherent QoS mechanism, we introduce QoS guarantees based on Weighted Fair Queuing (WFQ) scheme in accordance with the Max-Min Weighted Fair Share principle [48] for weight-based differentiation of users. WFQ is a discrete implementation of the generalized processor sharing (GPS) policy and an extension of fair queueing. It is realistically assumed that users have different bandwidth needs with varying priorities, therefore, all the ONUs do not request for an equal share of the resources at every given cycle. Consequently, allocating equal resources to them will lead to a waste of resources by the ONUs whose demands are lower than allocated grants, and some ONUs with higher requests will not be satisfied. Accordingly, some ONUs that have higher bandwidth demands are given more weight compared to ONUs with lower bandwidth demands and they are thus allocated relatively higher resources.

As shown in the pseudocode provided in Algorithm 2, we associate weights $\alpha_1, \alpha_2, ...,$ α_m with ONUs 1, 2, ..., m, which reflect their relative resource share. The resources are allocated to the ONUs in increasing order of their requests, normalized by their weights, with the small requests being fully granted first. In this case, the ONU with the lowest demand is maximized, if satisfied, only then the ONU with the second-lowest demand will be maximized. After the ONU with the second-lowest is satisfied, only then the ONU with the third-lowest demand will be maximized, and so on. Therefore, no ONU gets more than its demand, and the ONUs whose demands are not met get a fair share of the resources in proportion to their weights. This also avoids the situation where the resources will be monopolized by ONUs with bigger requests and consequently eliminating network congestion to some extent. We combine the WFQ principle with the LPT algorithm to give us WFQLPT, a hybrid algorithm that provides inherent QoS with the minimal makespan associated with LPT.

Algorithm 1. Pseudocode for LPT executed at the OLT for each cycle *i*.

Pseudocode of the LPT Heuristic Non-Preemptive Scheduler

for m = 1:M
if (ONU$_m$ is connected && Queue $\neq$ 0)
    Consider ONU;
    k = Connected_ONUs++;
  end if
end for
for m = 1: k
    sort J$_m$ in descending order based on their length;
    $L_r \geq L_s \geq L_t \geq \ldots \geq L_m$
end for
if ($\sum_0^k L_m \geq \delta_{max}$)
    δ_{max} is granted in cycle i;
    J$_m$ that are not assigned will be processed first in cycle i + 1
end if

Regarding the assignment of wavelength (ω), our proposed DBA algorithm combines the time allocation and wavelength algorithms following the JTWS scheme previously described in [12]. Once the OLT receives the requests from all connected ONUs by following an offline scheduling framework, it sorts the jobs according to the LPT scheme and thereafter assigns wavelengths in accordance with the Next Available Supported Channel (NASC) scheduling policy [33]. This allows the ONUs to be assigned to the next available wavelength, where their requests will be granted. The choice of NASC aligns with the principles of the LPT scheme, in which the unassigned task with the largest computation time is assigned to the next available wavelength [49].

The assignment of the wavelength according to NASC occurs in offline scheduling mode. The offline scheduling framework gives room for the LPT scheme and allows for applying WFQ QoS differentiation as scheduling decisions are made with full knowledge of all the jobs to be scheduled for a particular scheduling cycle. The cycle is the time difference between two consecutive allocation decisions. A profound advantage of the offline scheduling framework is the increased level of scheduling control, by which the OLT differentiates QoS. Specifically, the OLT adds all the ONUs with REPORT messages into a scheduling pool, and the scheduling is done after the OLT has sorted the REPORT messages and prioritized the ONUs based on their respective QoS. The channel is considered busy until the end of the last scheduled reservation, and then the procedure is applied for considering LTT when deciding whether or not to tune the supported wavelengths. Therefore, when a wavelength becomes free, it is assigned to the ONU with the longest job in the pool, as shown in Algorithm 3.

Our algorithm sorts the requests from the ONUs at the OLT according to the length of time needed for them to be processed, in descending order according to the LPT principle. The OLT sends grant messages (GATE) to the ONUs and schedules the ONU with the longest processing time first, which is then transmitted on the next available wavelength. We introduce the concept of LTT, and if the wavelength that the ONU is currently tuned to is the same as one that has been newly assigned by the OLT, then no laser tuning time is added. As shown in Algorithm 3, if the newly assigned wavelength is different from the current wavelength, the ONU checks the time needed for its current wavelength to become free and adds the laser tuning time to it. If the time needed to tune to a new wavelength is more, the ONU will remain on its current wavelength and no tuning time delay will be added. If the time to tune to a new wavelength is less, the ONU will tune to the newly assigned wavelength, and the tuning time delay will be added. This process happens continuously whenever GATE and REPORT messages are exchanged during the lifecycle of the communication between the OLT and the ONUs. Figure 1 illustrates the steps in the application of our algorithm.

Algorithm 2. Pseudocode for WFQ executed at the OLT for each cycle *i*.

Pseudocode of the Max-Min weighted fair-share queuing

---

for m = 1:M
   if (ONU$_m$ is connected && Queue $\neq$ 0)
      Consider ONU;
      k = Connected_ONUs++;
   end if
end for
for m = 1:k
      Search for the smallest weight (α_m) among the connected ONUs
      $\alpha_{min} = 1$;
      Normalize the remaining weights (α_m) of remaining connected ONUs based on
α ratio of normalization;
end for
for m = 1:k
      $\gamma = \sum_1^k \alpha_m$
end for
$\rho = \delta_{max} / \gamma$
for m = 1:k
      $\beta_m = \rho.\alpha_m$
end for
for m = 1:k
     if ($\beta_m \geq L_m$)
        Remaining += ($\beta_m - L_m$)
        Distribute the Remaining among unserved ONUs
     end if
     if (ONU$_m$ is not served)
        $\gamma = \sum_1^k \alpha_m$
        $\rho = \delta_{remaining} / \gamma$
        $\beta'_m = \rho.\alpha_m$
        $\beta'_m+ = \beta_m$
        if ($\sum_1^k \beta'_m \leq \delta_{max}$)
           grant all Jobs
        end if
        else
           The remaining jobs wait for the next cycle
        end else
     end if
end for
if ($\sum_1^k \beta_m \leq \delta_{max}$)
      grant all Jobs
end if
else
      The remaining jobs wait for the next cycle
end else

Algorithm 3. Pseudocode for NASC with LTT executed at the OLT for each cycle *i*.

Pseudocode of the Wavelength Assignment-NASC with LTT

---

if $(\omega_i = \omega_{i-1})$
    No tuning
    τ *is not considered*
    queue_delay $+= \theta(\omega, i)$
end if
else
        if$(\omega_i \neq \omega_{i-1})$
          if$(\tau \geq \theta(\omega, i-1))$
            No tuning
            τ *is not considered*
            $(\omega_i = \omega_{i-1})$
            queue_delay $+=\theta(\omega, i)$
          end if
        end if
        else
          Laser tunes
          τ *is considered*
          queue_delay $+= \tau$
        end else
end else

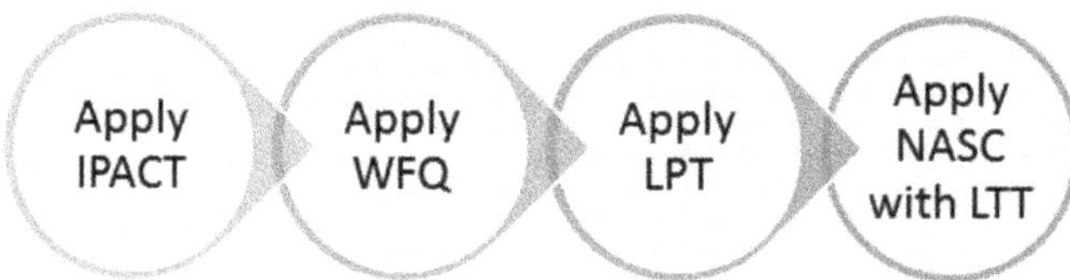

Figure 1. WFQLPT DBA algorithm sequence diagram.

4. Performance Evaluation

In this section, we evaluate the performance of our DBA algorithm. To validate the efficiency of our algorithm, we carried out extensive simulations using OPNET Modeller under different conditions.

4.1. Simulation Model

Our simulation setup consists of ONUs at the customers' premises, a centralized OLT, and an ODN that emulates a passive optical splitter/combiner, which splits the optical fiber cable running from the OLT to the ONUs. To check the impact on our algorithms from the number of ONUs in a PON system, we have created three different sets of scenarios, with 8 ONUs, 16 ONUs and 64 ONUs, respectively. The simulations involving 16 ONUs scenarios are further classified into two subcategories based on the distances from the ONUs to the OLT. One scenario is composed of 16 ONUs that are physically located at distances uniformly distributed between 18 km and 20 km, while the other set of 16 ONUs are physically located at distances uniformly distributed between 2 km and 20 km. In the downstream communication, the OLT broadcasts data to the ONUs, and each ONU filters the data sent to it and discards others. The upstream channel has a total capacity of 4 Gbps on four wavelengths, each one with a rate of 1 Gbps dynamically managed by the DBA. All ONUs are connected to their respective traffic sources and equipped with a packet generator over a link of 1 Gbps, thus avoiding possible bottlenecks. The maximum cycle time (δ_{max}) is 1 ms, and the sources generate self-similar traffic [43,50] with Hurst parameter H = 0.75 and a mean packet rate that is adjusted according to varying offered load. The frame size follows a uniform distribution with a lower limit of 512 bits and an upper limit of 12,144 bits, thus realistically modeling Ethernet traffic [43].

Several scenarios are created for the simulations in order to evaluate the effect on the algorithms from LPT scheduling, WFQ-based differentiation, and laser tuning time. A guaranteed weight of a specified percentage of the system's total bandwidth capacity is allocated to some ONUs, thus causing them to have different QoS. A laser tuning time of LTT = 10 μs is selected in reference to ITU-T G.989.2 specifications class 2 devices [27]. Traffic loads vary from 5% to 100% of the total load, where the maximum global offered load is 4 Gbps. The simulated algorithm sets are classified as IPACT, LPT, WFQ, and WFQLPT, depending on the configuration:

Set 1. IPACT with four wavelengths at LTT = 0 and 10 μs.
Set 2. LPT over IPACT with four wavelengths at LTT = 0 and 10 μs.
Set 3. WFQ with four wavelengths at LTT = 0 and 10 μs.
Set 4. LPT over WFQ (WFQLPT) with four wavelengths at LTT = 0 and 10 μs

For the simulations performed using the IPACT and LPT algorithms, the ONUs have an equal share of the total bandwidth and transmit on four wavelengths. To evaluate the QoS, we introduced different weights into the WFQ and WFQLPT algorithms. The 16 ONUs are distributed in such a way that ONU 1 and ONU 2 have a guaranteed share of, 20% (800 Mbps) and 10% (400 Mbps) respectively, and ONUs 3 to 16 each have 5% (200 Mbps).

4.2. Results

In terms of throughput and queue delay, we evaluate the performance of our novel DBA algorithm in comparison with IPACT, which has been extended to support four wavelengths. The results and discussion of each parameter based on allocated weight are presented as follows.

4.2.1. Throughput

The throughput represents the average number of bits per unit time, measured in Mbps, and it includes the Ethernet header (destination and source addresses) and trailer (frame check sequence) that are successfully transmitted by the ONUs. In this subsection, we present the comparative performance of the four DBA algorithms in terms of throughput for the upstream link under varying offered loads.

Figure 2 shows the QoS differentiation of our algorithms by allocating different weights to the ONUs in order to see the effect of LPT scheduling on the algorithms. We separate the Max-Min-based algorithms (WFQ and WFQLPT) from IPACT and LPT because of uneven bandwidth allocated to different ONUs. Figure 2 (left) presents the results for IPACT and LPT transmitting on all four wavelengths at 0 LTT for all the ONUs. In this case, all ONUs have an equal share of the system, with each ONU having a share of 250 Mbps. Here, we see that both IPACT and LPT behave similarly, as they can transmit an equal amount of throughput up to 220 Mbps before reaching saturation at an offered load of 210 Mbps. Thus, the introduction of LPT scheduling has no noticeable effect on IPACT in terms of throughput.

Figure 2 (right) shows the results for WFQ and WFQLPT with different weights. This scenario displays the QoS differentiation of the ONUs, with ONUs 1 and 2 having a share of 20% (800 Mbps) and 10% (400 Mbps), respectively; and ONUs 3 to 16 each have 5% (200 Mbps). The results show that ONU 1 can transmit up to 730 Mbps at a bandwidth efficiency of 91% before reaching saturation at an offered load of 725 Mbps. ONU 2 can transmit up to 375 Mbps at a bandwidth efficiency of 93.75% before reaching saturation at 360 Mbps. ONUs 3 to 16 can transmit up to 180 Mbps at a bandwidth efficiency of 90% while reaching saturation at an offered load of 185 Mbps. Thus, the introduction of LPT has no noticeable effect on the WFQ algorithm in terms of throughput.

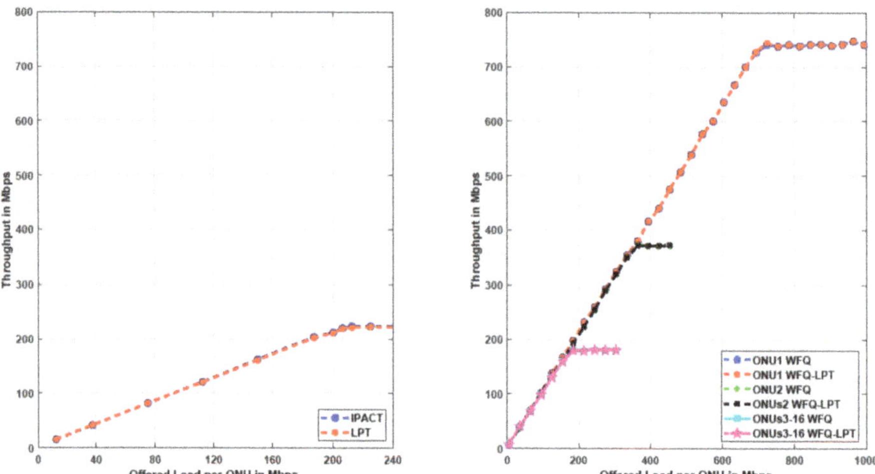

Figure 2. Throughput for all 16 ONUs at LTT = 0 µs; IPACT *vs.* LPT (**left**) and WFQ *vs.* WFQLPT (**right**).

The scenarios in Figure 3 show the effect of LTT on the throughput for the four algorithms. Figure 3 (left) displays the results for ONU1, whose share of the resources (250 Mbps) is equal to the remaining 15 ONUs under the IPACT and LPT algorithms at both LTT = 0 µs and LTT = 10 µs. With the ONUs transmitting on all four wavelengths, it can be seen that there is no noticeable difference in both IPACT and LPT for the two different LTTs (LTT = 0 µs and LTT = 10 µs), as the ONUs are transmitting at a maximum throughput of 225 Mbps. Thus, the laser tuning time does not affect the throughput when ONUs have an equal allocation of 250 Mbps.

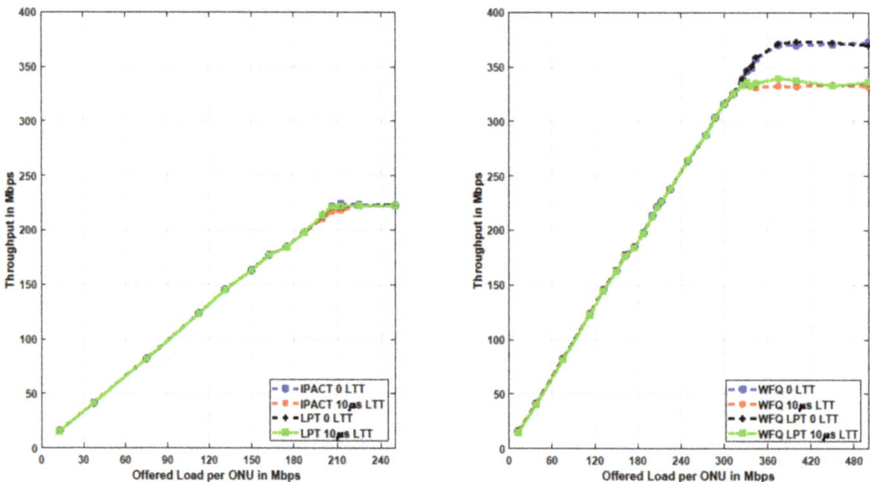

Figure 3. Throughput for ONU 1 at both LTT = 0 µs and LTT = 10 µs in a 16-ONU scenario; IPACT and LPT (**left**); and WFQ and WFQLPT (**right**).

In Figure 3 (right), we compare WFQ and WFQLPT at LTT = 0 µs and LTT = 10 µs for ONU1. In this case, ONU 1 has a share of 10% (400 Mbps) of the total resources, and the rest of the ONUs share the remaining 90%. We can see that at higher offered load, there is a difference in the throughput between LTT = 0 μs and LTT = 10 μs for both WFQ and WFQLPT. While ONU 1 can transmit up to 370 Mbps at LTT = 0 for both WFQ and WFQLPT, it is only able to transmit up to 335 Mbps when LTT of 10 μs is introduced before reaching saturation.

Figure 4 shows the results comparing the behavior of our system with different numbers of ONUs in the PON. We present the results for LPT and IPACT for 8 ONUs against 16 ONUs under LTT = 10 μs.

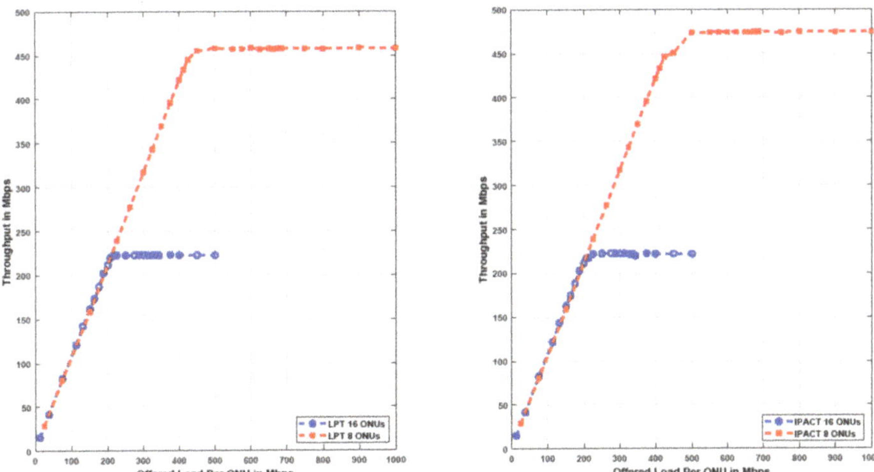

Figure 4. Throughput of all ONUs at LTT = 10 μs for 16 ONUs vs. 8 ONUs; LPT (**left**) and IPACT (**right**).

Figure 4 (left) shows that the throughput reaches 465 Mbps in the 8-ONU system and 225 Mbps in the 16-ONU system under the LPT algorithm. Figure 4 (right) shows that the throughput reaches 465 Mbps with 8 ONUs and 225 Mbps with 16 ONUs under the IPACT algorithm. These results show a proportional increase from 225 Mbps to about 465 Mbps when the ONUs decrease from 16 to 8. These results show that LPT and IPACT behave similarly, regardless of the number of ONUs in the PON system.

In order to check the effect of the distance between the OLT and ONUs in our algorithms, we compare the results for the set of 16 ONUs that are scattered within a distance range of 2–20 km versus those within a distance range of 18–20 km from the ONUs to the OLT. Figure 5 (left) shows the throughput in the case of the LPT algorithm. As we can see, LPT behaves the same within both ranges, as it can transmit up to 220 Mbps before reaching saturation at an offered load of 210 Mbps at LTT = 10 μs.

Figure 5 (right) shows the results for ONUs within 18–20 km against 2–20 km from the OLT under the IPACT algorithm at LTT = 10 μs. As we can see, the IPACT algorithm has the same behavior within both distance ranges, as the ONUs can transmit up to 222 Mbps before reaching saturation at 225 Mbps.

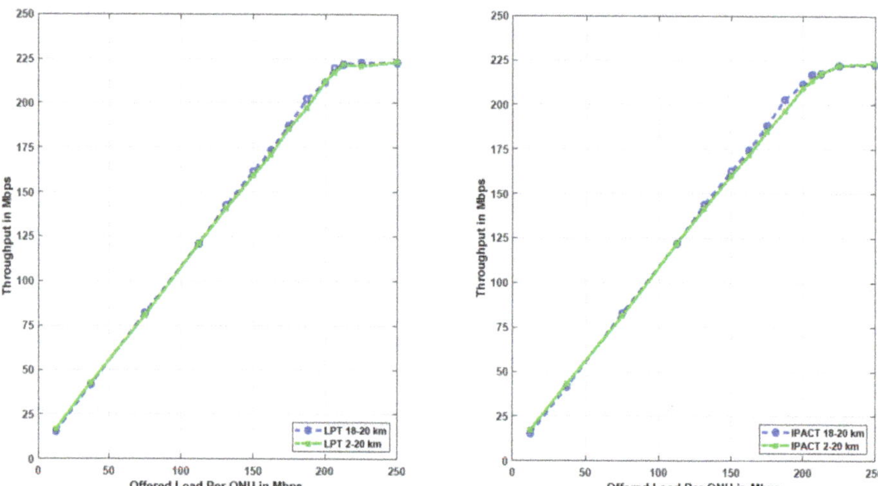

Figure 5. Throughput for all ONUs at LTT = 10 µs for a range of 18–20 km *vs.* 2–20 km for LPT (**left**) and for IPACT (**right**).

Figure 6 shows the LPT's impact on the throughput in function of the distance and load for LTT = 0 µs. ONU 1 and ONU 4 are located at 2 km and 20 km from the OLT, respectively. Figure 6 (left) shows the CDF of the throughput for IPACT. Therefore, we can conclude that the range in which the ONUs are spread has no impact on the throughput of the system at low loads. For both IPACT and LPT at heavy loads, the ONUs closest to the OLT have the same behavior. In contrast, for the distant ONUs, LPT suffers a deviation of less than 10% in terms of IPACT. This is because LPT reduces the delay of the frames even if the system works at heavy loads.

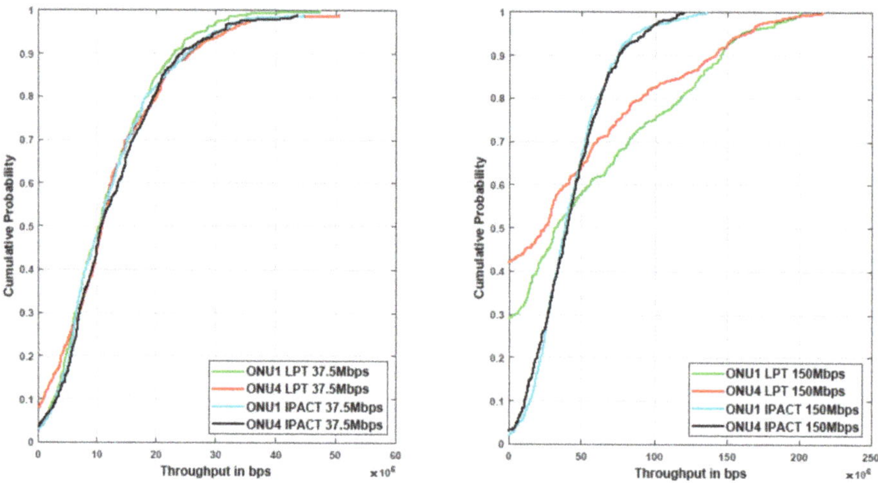

Figure 6. CDF of the throughput for ONU 1 and 4 at LTT = 0 µs; IPACT *vs.* LPT at low load (**left**) and heavy load (**right**).

4.2.2. Queue Delay

The queue delay measured in our simulations is the average packet waiting time in the ONU queues before being processed. The ONUs' queue delay is one of the components that forms the end-to-end delay, and the only one that is variable. In our scenario, packet transmission delay and propagation delay are negligible compared to the queue delay. We compare the queue delay for the IPACT, LPT, WFQ, and WFQLPT algorithms.

Figure 7 (left) shows the results for the average queue delay of all 16 ONUs for LPT versus IPACT at 0 LTT. In this scenario, all ONUs have an equal share of the system (250 Mbps) and transmit on all four wavelengths. The results show that the ONUs have a much lower delay of 0.08 ms under the LPT versus the IPACT algorithm (0.3 ms). It can be seen that ONUs under LPT and IPACT reach saturation at the same offered load of around 200 Mbps.

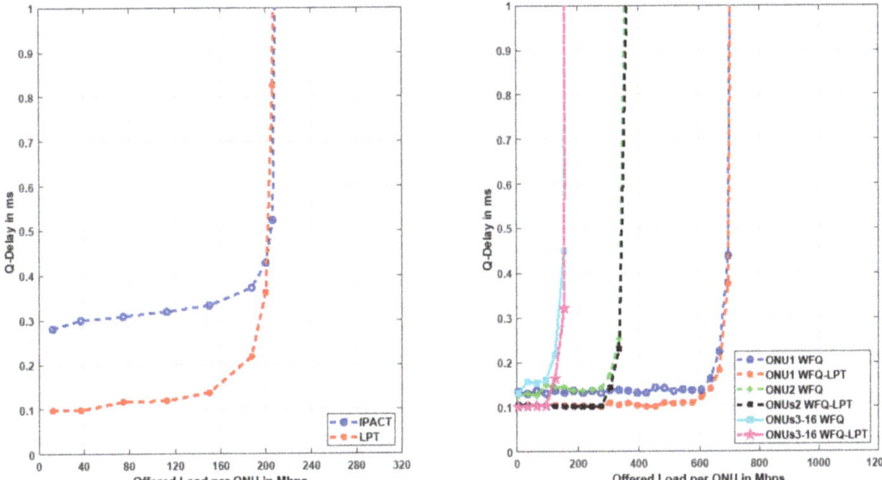

Figure 7. Queue delay for all ONUs at LTT = 0 μs; IPACT *vs.* LPT (**left**) and WFQ vs. WFQLPT (**right**).

Figure 7 (right) shows the results for ONUs under WFQ and WFQLPT using three different sets of weights (share of resources) for 0 LTT. ONU 1 has 800 Mbps, ONU 2 has about 400 Mbps, and ONUs 3 to 16 have 200 Mbps each. We can see that the ONUs have lower delays under WFQLPT (0.1 ms) than WFQ (0.15 ms). In addition, the variation in the allocated resources for the ONUs causes them to have three different sets of saturation points with respect to the allocated resources.

Figure 8 presents the effect of LTT on queue delay for the four algorithms compared to the offered loads. Figure 8 (left) shows the results for ONU 1 with an allocation of 250 Mbps under IPACT, and under LPT at both LTT = 0 μs and LTT = 10 μs. We can see that IPACT at LTT = 0 μs has a slightly lower queue delay (0.281 ms) than when LTT = 10 μs (0.296 ms).

In Figure 8 (right), WFQ and WFQLPT are compared under LTT = 0 μs and LTT = 10 μs. In this case, ONU 1 has an allocation of 400 Mbps. For WFQLPT, the effect of LTT can be seen, as the queue delay is slightly lower when LTT = 0 μs in the working area (0.279 ms at an offered load of 15 Mbps) than when LTT = 10 μs (0.297 ms at an offered load of 15 Mbps). If we compare the LPT queue delay with WFQLPT, we observe that the queue delay in the working area has increased by up to 50 μs. This increase is justified because the proposed WFQLPT algorithm offers quality of service, guarantees throughput, and thus minimizes delay. In the figure for WFQ with WFQLPT on the right, the former only guarantees throughput while the latter guarantees both parameters.

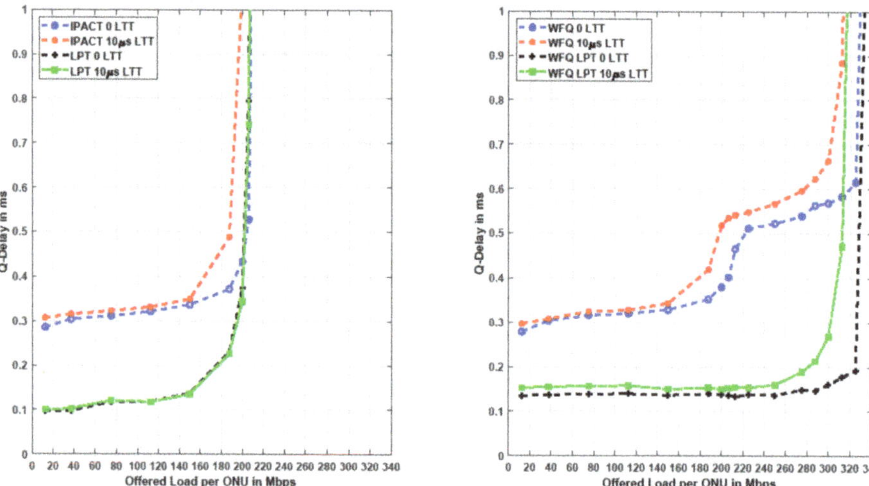

Figure 8. Queue delay for ONU 1 at both LTT = 0 μs and LTT = 10 μs; IPACT and LPT (**left**); and WFQ and WFQLPT (**right**).

Figure 9 shows the results for the simulations of the IPACT and LPT algorithms, which were run to check the behavior of our system when different numbers of ONUs are connected. We compare the queue delay for the three scenarios with 64 ONUs vs. 16 ONUs vs. 8 ONUs at LTT = 10 μs. The results show that, apart from the expected rescaling of the saturation point, the number of ONUs does not have any important impact, and the LPT and IPACT algorithms behave similarly. Figure 9 (left) shows that the queue delay under the LPT algorithm is kept at almost the same value between 0.1 ms and 0.15 ms in the three scenarios until it reaches saturation at an offered load of approximately 50Mbps, in the case of 64 ONUs, 200 Mbps, in the case of 16 ONUs, and of 400 Mbps for the 8 ONUs.

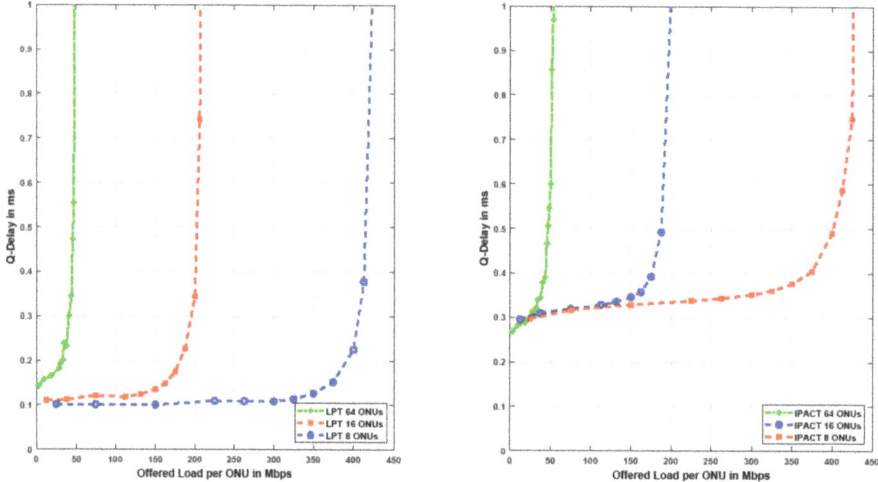

Figure 9. Average queue delay for all ONUs at LTT = 10 μs for 64 ONUs, 16 ONUs and 8 ONUs; LPT (**left**) and IPACT (**right**).

Figure 9 (right) shows that, under the IPACT algorithm, the queue delay is similar for the three scenarios, as it is kept at around 0.3 ms until reaching an offered load of 48 Mbps in the case of 64 ONUs, 190 Mbps, in the case of 16 ONUs, and of 400 Mbps with 8 ONUs.

These results confirm that the three sets of ONUs (64, 16, and 8) have similar behaviors under the IPACT and LPT algorithms.

In Figure 10, we evaluate the performance of the LPT and IPACT algorithms in two scenarios with 16 ONUs: in the first, the ONUs are scattered within a distance range of 2–20 km to the OLT; and in the second, the distance range is 18–20 km. Figure 10 (left) shows the queue delay results for the LPT algorithm, and it is evident that the behavior is the same in both ranges, as the queue delay is kept at 0.11 ms before reaching saturation at 190 Mbps.

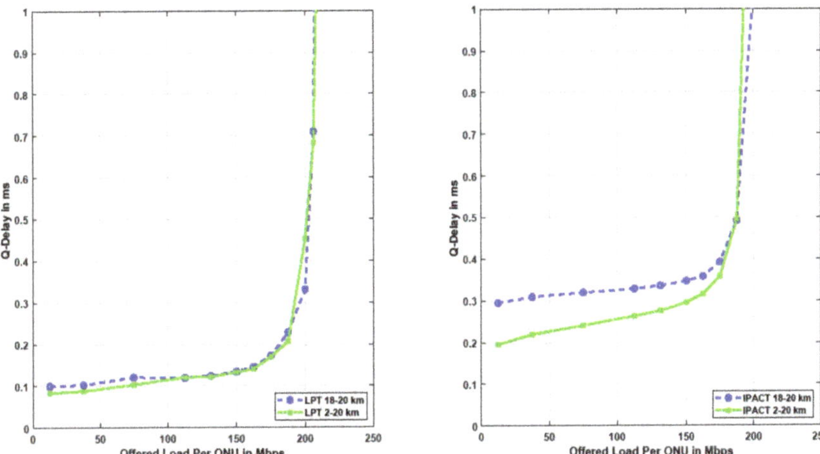

Figure 10. Queue delay for a 16-ONU system with a distance range of 2–20 km vs. 18–20 km at LTT = 10 µs; LPT (**left**) and IPACT (**right**).

Figure 10 (right) shows the IPACT algorithm's queue delay results for the set of ONUs within 2–20 km vs. 18–20 km. It can be deduced that the queue delay in the 18–20 km range goes from 0.3 ms until reaching 0.35 ms near the saturation point at offered loads of 190 Mbps. Meanwhile, in the 2–20 km range scenario, the queue delay goes from 0.2 ms to 0.3 ms near the saturation point at offered loads of 190 Mbps.

Figure 11 shows the Cumulative Distribution Function (CDF) of the queue delay for ONU 1 and ONU 4 at offered loads of 37.5 Mbps and 150 Mbps under the IPACT and LPT algorithms for LTT = 10 µs. ONU 1 and ONU4 are located at, respectively, 2.61 km and 17.43 km from the OLT in the 16-ONU scenario. Figure 11 (left) shows that, at low load, LPT's margin of improvement is between 50 µs to 100 µs for IPACT, while under LPT the impact of the distance is limited to below 25 µs.

In Figure 11 (right), we show the queue delay for both ONU 1 and ONU 4 at an offered load of 150 Mbps. For heavy loads, the improvement of LPT with respect to IPACT is more evident, as it increases the difference by 150 µs. Furthermore, the variability of the queue delay for heavy loads and different distances is kept at around 40 µs. This delay is similar to that of low loads. Thus, it can be deduced from Figures 10 and 11 that even though the queue delay increases together with the distance between the ONU and OLT, the difference in the queue delay narrows as the offered load increases.

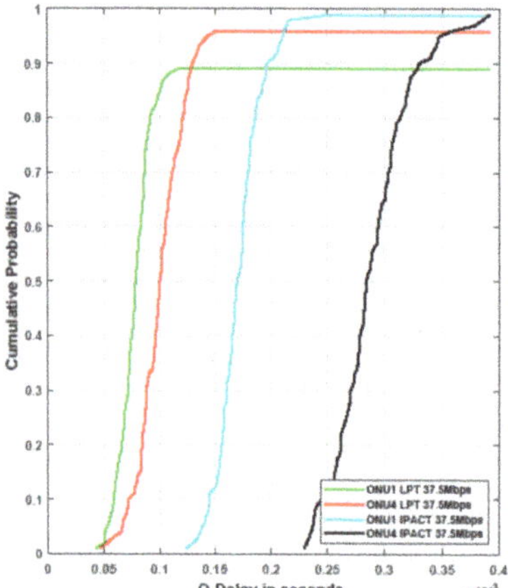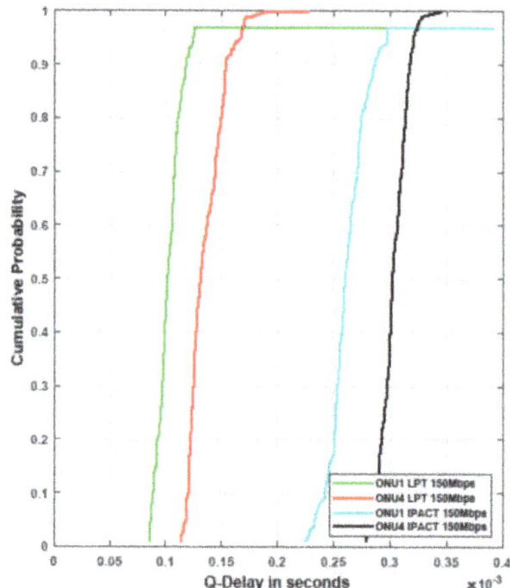

Figure 11. CDF of the queue delay for ONU 1 and ONU 4 under the IPACT and LPT algorithms for LTT = 10 µs at offered loads of 37.5 Mbps (**left**) and 150 Mbps (**right**).

4.3. Discussion of the Results

The results show four main aspects. First, we have our analysis on the influence of laser tuning time on IPACT, in terms of throughput, and queue delay as a function of the system load and the distance between the ONUs and the OLT. Second, the LPT optimizes the queue delay more than IPACT in all the above scenarios. Third, as the WFQ guarantees a throughput for each user, we have evaluated the impact when the LTT is introduced. Finally, the performance of the WFQLPT has been evaluated, which guarantees a minimum queue delay and the bandwidth requested by the different ONUs.

We can achieve an average bandwidth efficiency of about 85% in the upstream link, which conforms to the minimum absolute efficiency that is stipulated in [51]. The inefficiency in the system is a result of overhead from encapsulation, such as control message overhead (which represents bandwidth lost to GATE and REPORT message exchanges between the ONUs and OLT), guard band overhead, discovery overhead, and frame delineation overhead [51]. The overhead consumes up to 16.37% of the bandwidth, with the minimum throughput being 836.3 Mbps on a 1 Gbps link.

Introducing a realistic LTT of 10 µs provokes a noticeable decrease in the throughput and an increase in the queue delay. The effect on the throughput is seen only at higher loads, above 320 Mbps (80% offered), with the delay introduced as a result of LTT reducing the throughput capability of the ONUs by 10% compared to when there is no LTT, thus reducing the bandwidth utilization of the system. The effect of LTT is not noticed at lower loads because the system is not operating at nearly full capacity. As such, the system can transmit up to the maximum allowable throughput. The effect of LTT on the queue delay is seen only at the point where the queue delay starts skyrocketing, which is at a much higher offered load when the LTT is 0 than when the LTT is 10 µs. At lower loads, the queue delay is kept minimal and comparable to when there is no LTT applied.

The number of ONUs connected to the OLT does not in any way affect the performance of the DBA algorithms in terms of queue delay and throughput. When the resources are to be shared equally among the ONUs in the PON, the algorithms behave the same way, with each ONU's throughput and queue delay having similar values. The distance between the ONU and the OLT within the maximum allowed distance (20 km) of PON does not have any impact on the throughput. It does not matter where the ONU is located in the PON, the throughput will still be the same. However, the queue delay is affected and it decreases as the ONUs become closer to the OLT. However, the impact of the distance between the ONU and the OLT decreases as the offered load increases.

These results emphasize the need to consider LTT when designing DBA and evaluating its performance in order to obtain realistic results and model the behavior of a system whose delay requirements are within a band of 1 ms to 100 ms [52], which is what critical services demand in 5G networks.

The LPT scheme is introduced to solve the problem of minimizing the total finish time when scheduling requests on multiple wavelengths. When LPT is applied to the IPACT algorithm, the queue delay is reduced (by 73%), but there is no noticeable effect on the throughput. Applying LPT to the WFQ algorithm gives us the WFQLPT algorithm, which is a hybrid that combines the low delay of LPT with the QoS differentiation provided by WFQ. Therefore, WFQLPT achieves QoS differentiation and proves to be superior to WFQ in terms of delay, which is reduced by approximately 33%. In terms of throughput, there is no noticeable difference between WFQLPT and WFQ.

When implementing QoS differentiation in WFQ and WFQLPT algorithms, the ONUs with higher priority obtain their share and the remaining ONUs obtain a fair share of the resources without any of them being starved. The introduction of QoS based on WFQ reduces wasted bandwidth because the bandwidth utilization is in the region of 91%, which is higher than when no QoS is applied (88%), and there is no noticeable impact on the delay.

5. Conclusions

TWDM-based PON is a promising technology with great potential for providing the high bandwidth capacity and low latency required by emerging services. TWDM-PON DBA algorithms need to take into account the laser tuning time (LTT), which is often ignored. In this paper, we have presented a WFQLPT, QoS-aware algorithm that considers LTT. Our algorithm builds on IPACT by adding the capability of supporting four wavelengths. We apply a scheduling mechanism based on the LPT scheme that arranges the requests from the ONUs in descending order before being scheduled on the assigned wavelengths in accordance with the NASC principle, thus reducing delay. As IPACT is known to lack QoS capability, we have introduced weight-based QoS differentiation based on Max-Min Weighted Fair Share in order to ensure fair sharing of resources. We evaluated our approach through simulations, and our results show that the bandwidth is shared fairly among the users while wavelengths are allocated in a more balanced manner. Introducing WFQ guarantees the allocation of resources based on the Service Level Agreement (SLA) while keeping delay bounded. We can see the effect of LPT in reducing the average packet delay on IPACT by 73% and on the WFQ algorithm by 33%. We have also shown that the delay introduced as a result of LTT gives the system more realistic behavior in terms of the throughput and queue delay.

This paper opens up a new horizon of research on the implementation of the DBA algorithm while focusing on efficient energy utilization in order to save power, which is a worthwhile contribution, given the predominant role of PONs in next-generation networks. We plan to introduce power-saving features such as laser doze/sleep mode [53] and further exploit the laser tuning time to achieve optimal results while keeping the delay minimal. We also intend to enhance the algorithm by implementing it in a just-in-time manner in order to further reduce the delay [33]. Furthermore, we will work on improving the management architecture of TWDM-PON by introducing the software-defined networking principle to decouple the OLT and move the DBA functions to a centralized controller, which will thus manage the network with flexibility [54]. An interesting direction for future research will be to consider Long-Reach Passive Optical Networks (LRPON), with a multi-thread polling scheme to enhance their operations.

Author Contributions: Conceptualization, A.H. and M.Z.; methodology, A.H. and M.Z.; software, A.H., M.Z. and A.B.; validation, D.R., S.S., and J.R.P.; formal analysis, A.H., and M.Z.; investigation, A.H. and M.Z.; resources, A.H. and M.Z.; writing—original draft preparation, A.H. and M.Z.; writing—review and editing, A.H., M.Z., D.R., S.S. and J.R.P.; visualization, A.H., M.Z., D.R., S.S. and J.R.P.; supervision, D.R., S.S., J.R.P., and A.B. All authors have read and agreed to the published version of the manuscript.

Funding: This work has been supported by the Agencia Estatal de Investigación of Spain under project PID2019-108713RB-C51/AEI/10.13039/501100011033.

Data Availability Statement: Not applicable.

Conflicts of Interest: The authors declare no conflict of interest.

References

1. Taylor, R.; Baron, D.; Schmidt, D. The world in 2025—Predictions for the next ten years. In Proceedings of the 10th International Microsystems, Packaging, Assembly and Circuits Technology Conference (IMPACT), Taipei, Taiwan, 21–23 October 2015; pp. 192–195.
2. Hadi, M.; Bhar, C.; Agrell, E. General QoS-aware scheduling procedure for passive optical networks. *J. Opt. Commun. Netw.* **2020**, *12*, 217–226. [CrossRef]
3. Pizzinat, A.; Chanclou, P.; Saliou, F.; Diallo, T. Things You Should Know About Fronthaul. *J. Light. Technol.* **2015**, *33*, 1077–1083. [CrossRef]
4. Nakayama, Y.; Hisano, D. Wavelength and Bandwidth Allocation for Mobile Fronthaul in TWDM-PON. *IEEE Trans. Commun.* **2019**, *67*, 7642–7655. [CrossRef]
5. Kondepu, K.; Valcarenghi, L.; Castoldi, P. Balancing the impact of ONU tuning overhead in reconfigurable TWDM-PONs: An FPGA-based evaluation. In Proceedings of the IEEE Global Communications Conference (GLOBECOM), San Diego, CA, USA, 6–10 December 2015; pp. 1–6.
6. Memon, K.A.; Mohammadani, K.H.; Shaikh, A.; Ullah, S.; Zhang, Q.; Das, B.; Ullah, R.; Tian, F.; Xin, X. Demand forecasting DBA algorithm for reducing packet delay with efficient bandwidth allocation in XG-PO. *Electronics* **2019**, *8*, 147. [CrossRef]
7. Ruffini, M.; Achouche, M.; Arbelaez, A.; Bonk, R.; Di Giglio, A.; Doran, N.J.; Furdek, M.; Jensen, R.; Montalvo, J.; Parsons, N.; et al. Access and Metro Network Convergence for Flexible End-to-End Network Design. *J. Opt. Commun. Netw.* **2017**, *9*, 524–535. [CrossRef]
8. Rafiq, A.; Hayat, M.F. QoS-Based DWBA Algorithm for NG-EPON. *Electronics* **2019**, *8*, 230. [CrossRef]
9. Liu, X.; Deng, N.; Zhou, M.; Wang, Y.; Tao, M.; Zhou, L.; Li, S.; Zeng, H.; Megeed, S.; Shen, A.; et al. Enabling Technologies for 5G-Oriented Optical Networks. In Proceedings of the Optical Fiber Communications Conference and Exhibition (OFC), San Diego, CA, USA, 3–7 March 2019; pp. 1–3.
10. Afraz, N.; Slyne, F.; Gill, H.; Ruffini, M. Evolution of Access Network Sharing and Its Role in 5G Networks. *Appl. Sci.* **2019**, *9*, 4566. [CrossRef]
11. Nakayama, Y.; Uzawa, H.; Hisano, D.; Ujikawa, H.; Nakamura, H.; Terada, J.; Otaka, A. Efficient DWBA Algorithm for TWDM-PON with Mobile Fronthaul in 5G Networks. In Proceedings of the IEEE Global Communications Conference (GLOBECOM), Singapore, 4–8 December 2017; pp. 1–6.
12. Dixit, A.; Lannoo, B.; Colle, D.; Pickavet, M.; Demeester, P. Dynamic bandwidth allocation with optimal wavelength switching in TWDM-PONs. In Proceedings of the 15th International Conference on Transparent Optical Networks (ICTON), Cartagena, Spain, 23–27 June 2013; pp. 1–4.
13. Park, C.; Min, J.; Kim, S. Semi-passive optical front-haul supporting channel monitoring and link protection for the cloud radio access network. *IET Commun.* **2019**, *13*, 442–448. [CrossRef]
14. Xue, L.; Yi, L.; Ji, H.; Li, P.; Hu, W. Symmetric 100-Gb/s TWDM-PON in O-Band Based on 10G-Class Optical Devices Enabled by Dispersion-Supported Equalization. *J. Light. Technol.* **2017**, *36*, 580–586. [CrossRef]
15. Jaffer, S.S.; Hussain, A.; Qureshi, M.A.; Khawaja, W.S. Towards the Shifting of 5G Front Haul Traffic on Passive Optical Network. *Wirel. Pers. Commun.* **2020**, *112*, 1549–1568. [CrossRef]
16. Liem, A.T.; Hwang, I.-S.; Nikoukar, A.; Pakpahan, A.F. SD-Enabled Mobile Fronthaul Dynamic Bandwidth and Wave-length Allocation (DBWA) Mechanism in Converged TWDM-EPON Architecture. In Proceedings of the 6th International Conference on Cyber and IT Service Management (CITSM), Parapat, Indonesia, 7–9 August 2018; pp. 1–6.
17. Zhang, J.; Ansari, N. Scheduling Hybrid WDM/TDM Passive Optical Networks with Nonzero Laser Tuning Time. *IEEE/ACM Trans. Netw.* **2010**, *19*, 1014–1027. [CrossRef]
18. Buttaboni, A.; de Andrade, M.; Tornatore, M.; Pattavina, A. Dynamic bandwidth and wavelength allocation with coexist-ing transceiver technology in WDM/TDM PONs. *Opt. Switch. Netw.* **2016**, *21*, 31–42. [CrossRef]
19. Ojeyinka, T.O. Bin packing algorithms with applications to passenger bus loading and multiprocessor scheduling problems. *Commun. Appl. Electron.* **2015**, *2*, 38–44.
20. Marbach, P. Priority service and max-min fairness. In Proceedings of the Twenty-First Annual Joint Conference of the IEEE Computer and Communications Societies, New York, NY, USA, 23–27 June 2002; Volume 1, pp. 266–275.
21. Kramer, G.; Mukherjee, B.; Pesavento, G. Interleaved Polling with Adaptive Cycle Time (IPACT): A Dynamic Bandwidth Distribution Scheme in an Optical Access Network. *Photon. Netw. Commun.* **2002**, *4*, 89–107. [CrossRef]
22. Horvath, T.; Munster, P.; Vojtech, J. Deployment of PON in Europe and Deep Data Analysis of GPON. *Telecommun. Syst. Princ. Appl. Wirel. Opt. Technol.* **2019**, 56–76. [CrossRef]
23. Law, D.; Dove, D.; D'Ambrosia, J.; Hajduczenia, M.; Laubach, M.; Carlson, S. Evolution of ethernet standards in the IEEE 802.3 working group. *IEEE Commun. Mag.* **2013**, *51*, 88–96. [CrossRef]

24. IEEE Standards Association. IEEE Standard for Ethernet, IEEE Std 802.3-2018 (Revision of IEEE Std 802.3-2015), IEEE. 31 August 2018. Available online: https://ieeexplore.ieee.org/document/8457469 (accessed on 29 April 2021).

25. ITU-T Recommendation, G.9807.1, 10-Gigabit-Capable Symmetric Passive Optical Network (XGS-PON), International Telecommunications Union. 1 June 2016. Available online: https://www.itu.int/rec/T-REC-G.9807.1/en (accessed on 29 April 2021).

26. IEEE Standards Association, 802.3ca-2020 Approved Draft Standard for Ethernet Amendment, IEEE. June 2020. Available online: https://standards.ieee.org/standard/802_3ca-2020.html (accessed on 24 June 2020).

27. ITU-T Recommendation, G.989.2, 40-Gigabit-Capable Passive Optical Networks 2 (NG-PON2): Physical Media Dependent (PMD) Layer Specification, International Telecommunications Union. February 2019. Available online: https://www.itu.int/rec/T-REC-G.989.2 (accessed on 29 April 2021).

28. Wey, J.S.; Nesset, D.; Valvo, M.; Grobe, K.; Roberts, H.; Luo, Y.; Smith, J. Physical layer aspects of NG-PON2 standards—Part 1: Optical link design. *IEEE/OSA J. Opt. Commun. Netw.* **2016**, *8*, 33–42. [CrossRef]

29. ITU-T Recommendation G.984.6, Gigabit-Capable Passive Optical Networks (GPON): Reach Extension, International Telecommunications Union. March 2008. Available online: https://www.itu.int/rec/T-REC-G.984.6-200803-I/en (accessed on 29 April 2021).

30. ITU-T Recommendation G.9807.2, 10 Gigabit-Capable Passive Optical Networks (XG(S)-PON): Reach Extension, International Telecommunications Union. August 2017. Available online: https://www.itu.int/rec/T-REC-G.9807.2/en (accessed on 29 April 2021).

31. ITU-T Work Programme, Work Item G.9807.3 (ex G.SuperPON). Available online: https://www.itu.int/ITU-T/workprog/wp_item.aspx?isnD15208 (accessed on 29 April 2021).

32. DeSanti, C.; Du, L.; Guarin, J.; Bone, J.; Lam, C. Super-PON: An evolution of access networks. *Opt. Commun. Netw.* **2020**, *12*, D66–D76. [CrossRef]

33. McGarry, M.P.; Reisslein, M.; Colbourn, C.J.; Maier, M.; Aurzada, F.; Scheutzow, M. Just-in-time scheduling for multichan-nel EPON. *Lightwave Technol.* **2008**, *26*, 1204–1216. [CrossRef]

34. Dixit, A.; Lannoo, B.; Colle, D.; Pickavet, M.; Demeester, P.; Abhishek, D. Energy efficient DBA algorithms for TWDM-PONs. In Proceedings of the 17th International Conference on Transparent Optical Networks (ICTON), Budapest, Hungary, 5–9 July 2015; pp. 1–5.

35. Kanonakis, K.; Tomkos, I. Improving the efficiency of online upstream scheduling and wavelength assignment in hybrid WDM/TDMA EPON networks. *IEEE J. Sel. Areas Commun.* **2010**, *28*, 838–848. [CrossRef]

36. Dixit, A.; Colle, D.; Lannoo, B.; Demeester, P.; Pickavet, M. Novel DBA algorithm for energy efficiency in TWDM-PONs. In Proceedings of the 9th European Conference and Exhibition on Optical Communication (ECOC), London, UK, 22–26 September 2013; pp. 1–3.

37. Wang, W.; Guo, W.; Hu, W. Adaptive wavelength allocation pattern for an online DWBA in the NG-EPON. *OSA Contin.* **2018**, *1*, 690–702. [CrossRef]

38. Zaouga, A.; De Sousa, A.; Najja, M.; Monteiro, P. Low Latency Dynamic Bandwidth Allocation Algorithms for NG-PON2 to Support 5G Fronthaul and Data Services. In Proceedings of the 21st International Conference on Transparent Optical Networks (ICTON), Angers, France, 9–13 July 2019; pp. 1–4.

39. Wang, H.; Su, S.; Gu, R.; Ji, Y. A minimum wavelength tuning scheme for dynamic wavelength assignment in TWDM-PON. In Proceedings of the 7th IEEE International Conference on Software Engineering and Service Science (ICSESS), Beijing, China, 26–28 August 2016; pp. 1–3.

40. Zhang, H.; Zhang, M.; Liu, X.; Wang, D.; Jiang, L. A Multi-OLTs and Virtual Passive Optical Network for Hybrid Net-work. In Proceedings of the 7th IEEE International Conference on Software Engineering and Service Science (ICSESS), Beijing, China, 26–28 August 2016; pp. 1009–1012.

41. Zhang, L.; Qi, J.; Wei, K.; Zhang, W.; Feng, Y.; Hou, W. High-priority first dynamic wavelength and bandwidth allocation algorithm in TWDM-PON. *Opt. Fiber Technol.* **2019**, *48*, 165–172. [CrossRef]

42. Gravalos, I.; Yiannopoulos, K.; Papadimitriou, G.; Varvarigos, E.A. A modified max-min fair dynamic bandwidth alloca-tion algorithm for XG-PONs. In Proceedings of the 19th European Conference on Networks and Optical Communications (NOC), Milan, Italy, 4–6 June 2014; pp. 57–62.

43. Navarro, M.C. Study and Implementation of New DBA for WDM-PON. Master's Thesis, Universitat Politècnica de Cata-Lunya (UPC), Barcelona, Spain, 2010.

44. Agrawal, T.K.; Sahu, A.; Ghose, M.; Sharma, R. Scheduling chained multiprocessor tasks onto large multiprocessor system. *Computing* **2017**, *99*, 1007–1028. [CrossRef]

45. Grigoriu, L. Approximation for Scheduling on Parallel Machines with Fixed Jobs or Unavailability Periods. *Sched. Probl. New Appl. Trends* **2020**, 1–17. [CrossRef]

46. Lee, C.-Y.; Massey, J.D. Multiprocessor scheduling: Combining LPT and MULTIFIT. *Discret. Appl. Math.* **1988**, *20*, 233–242. [CrossRef]

47. Pinedo, M.; Hadavi, K. Scheduling: Theory, Algorithms and Systems Development. In *Operations Research Proceedings 1991*; Springer: Berlin/Heidelberg, Germany, 1992; pp. 35–42.

48. Li, G.; Qian, Y.; Yang, Y.R. On max-min fair allocation for multi-source transmission. *ACM Sigcomm Comput. Commun. Rev.* **2019**, *48*, 2–8. [CrossRef]
49. Alkallak, I.N. A Modified for Largest Processing Time Scheduling Algorithm in Multiprocessor. *Tikrit J. Pure Sci.* **2011**, *16*, 280–283.
50. Melo, E.F.; de Oliveira, H.M. An Overview of Self-Similar Traffic: Its Implications in the Network Design. *Rev. Tecnol. Inf. E Comun.* **2020**, *9*, 38–46.
51. Kramer, G. How Efficient Is EPON? White Paper. Available online: https://www.ieee802.org/3/efm/public/p2mp_email/pdf0 0001.pdf (accessed on 29 April 2021).
52. Parvez, I.; Rahmati, A.; Guvenc, I.; Sarwat, A.I.; Dai, H. A Survey on Low Latency Towards 5G: RAN, Core Network and Caching Solutions. *IEEE Commun. Surv. Tutor.* **2018**, *20*, 3098–3130. [CrossRef]
53. Khalili, H.; Rincón, D.; Sallent, S.; Piney, J.R. An Energy-Efficient Distributed Dynamic Bandwidth Allocation Algorithm for Passive Optical Access Networks. *Sustainability* **2020**, *12*, 2264. [CrossRef]
54. Zehri, M.; Haastrup, A.; Rincón, D.; Piney, J.R.; Sallent, S.; Bazzi, A. Leveraging SDN-Based Management for Improved Traffic Scheduling in PONs. In Proceedings of the 22nd International Conference on Transparent Optical Networks (ICTON), Bari, Italy, 19–23 July 2020; pp. 1–4.

Article

Design of a Free Space Optical Communication System for an Unmanned Aerial Vehicle Command and Control Link

Yiqing Zhang [1,2], Yuehui Wang [1], Yangyang Deng [3], Axin Du [1] and Jianguo Liu [1,*]

[1] State Key Laboratory of Integrated Optoelectronics, Institute of Semiconductors, Chinese Academy of Sciences, Beijing 100083, China; zyq0806@semi.ac.cn (Y.Z.); www@semi.ac.cn (Y.W.); duaxin@semi.ac.cn (A.D.)

[2] School of Electronic, Electrical and Communication Engineering, University of Chinese Academy of Sciences, Beijing 100049, China

[3] School of Microelectronics, University of Chinese Academy of Sciences, Beijing 100049, China; dengyangyang18@mails.ucas.ac.cn

* Correspondence: jgliu@semi.ac.cn

Abstract: An electromagnetic immune Free Space Optical Communication (FSOC) system for an Unmanned Aerial Vehicle (UAV) command and control link is introduced in this paper. The system uses the scheme of omnidirectional receiving and ground scanning transmitting. It has a strong anti-turbulence ability by using a large area detector and short-focus lens. The design of omnidirectional communication improves the ability of anti-vibration and link establishment. Pure static reception has no momentum effect on the platform. The receiver is miniaturized under no use of a gimbal mirror system, beacon camera system, Four-Quadrant Photodetector (QPD) and multi-level lens system. The system can realize omnidirectional reception and the communication probability in 1 s is greater than 99.99%. This design strengthens the ability of the FSOC system, so it can be applied in the UAV command and control, the satellite submarine communication and other occasions where the size of the platform is restricted.

Keywords: FSOC system; omnidirectional communication; miniaturization

Citation: Zhang, Y.; Wang, Y.; Deng, Y.; Du, A.; Liu, J. Design of a Free Space Optical Communication System for an Unmanned Aerial Vehicle Command and Control Link. *Photonics* **2021**, *8*, 163. https://doi.org/10.3390/photonics8050163

Received: 18 March 2021
Accepted: 8 May 2021
Published: 14 May 2021

1. Introduction

Compared with traditional optical fiber communication and microwave communication, FSOC has the advantages of high communication speed, strong anti-interference ability, high security and small size [1,2]. It has many applications in the civil emergency information transmission, the information security transmission under electromagnetic interference and the space information network construction [3–5]. Related researches have been carried out in References [6–16]. Besides, UAV has the advantage of flexible deployment and is developing rapidly in the direction of miniaturization and high speed [17,18]. However, UAV command and control systems are facing challenges on interference made by a Wireless Fidelity (Wi-Fi) signal and its own interference made to other facilities, typically the influence in the airport. Using the FSOC system on the UAV helps enhance the communication ability and eliminate the strict condition on the electromagnetic environment, which brings huge impact to the UAV's communication mode [18].

The FSOC system applied in UAV also faces many challenges. First, FSOC inevitably faces the problem of atmospheric turbulence, and UAV produces serious atmospheric turbulence in the flight process, which significantly damages the communication probability. Second, if the UAV loses its tracking during flight, it is difficult for normal FSOC equipment to complete the link establishing in a short time, which may cause the UAV to go out of control. Third, the moment of inertia generated by the rotation of the FSOC system during the search process could disturb the flight of UAVs.

Many methods have been proposed to apply FSOC equipment on UAV. In Reference [17], the integrated optical module helps to support both the signal beam and the beacon beam. This scheme can reduce volume to a certain extent, but the static transmission distance is only 50 m. In Reference [19] and Reference [20], adaptive optics technology is used to compensate the atmospheric disturbance in real time. They realize the demonstration experiment of low-power and high-speed air to ground laser communication. However, the equipment is so heavy that can only be carried by P68 aircraft, which can carry 680 kg. A precious tracking system based on fast steering mirror is adopted in Reference [21]. The scheme also reduces the intensity jitter at the cost of increasing the volume and complexity of the receiver, but it cannot work in the vibration environment because of the poor anti-vibration performance of the fast steering mirror.

In this paper, a scheme of omni-directional receiving is proposed, which can realize anti-jamming UAV command and control. It has the following advantages:

1. Large area detectors and short focus lenses are used at the receiving end, which greatly improve the anti-turbulence ability of the system;
2. An omni-directional receiving communication mode is adopted, thus enhancing its abilities for anti-vibration and link establishing;
3. The receiver adopts the pure static receiving mode, which does not affect the momentum of the platform and help the UAV free from flying interference;
4. The receiver of the UAV can be miniaturized for it eliminates the gimbal mirror system, the beacon camera system and the QPD or the multi-level lens system.

Because the command and control center on the ground has no need to be miniaturized, it can use the gimbal mirror system and some other large devices to transmit the command and control information directionally. Moreover, because the communication is a one-way link, no receiving device in required on the ground, as shown in Figure 1. Therefore, the design of the command and control center on the ground is not discussed in detail here.

Figure 1. One-way communication link.

2. Principle

The communication performance of the system is analyzed in this section. Reference [22] calculates the outage probability in stratospheric FSOC system by Rician factor, inter-high-attitude platform distance and misalignment-induced fading. Since the UAV command and control does not need continuous communication, the connectivity probability within 1 s is calculated to characterize the communication quality. When the Gaussian beam transmitting in the turbulence, the irradiance strength at receiver is not a fixed value. The irradiance probability distribution function (PDF) is generally accepted to be lognormal, and takes the form of [23]:

$$p(I) = \frac{1}{I\sigma_1(r,L)\sqrt{2\pi}} \exp\left\{-\frac{\left[\ln\left(\frac{I}{I_0\langle I(r,L)\rangle}\right) + \frac{1}{2}\sigma_1^2(r,L)\right]^2}{2\sigma_1^2(r,L)}\right\}, \tag{1}$$

where $\sigma_1(r,L)$ is the scintillation index of Gaussian beam and $\langle I(r,L)\rangle$ is the irradiance of beam in normalized radiation medium, and the $I_0\langle I(r,L)\rangle$ is the mean irradiance. The mean irradiance of Gaussian beam takes the form of:

$$\langle I(r,L)\rangle = \frac{W_0^2}{W^2}\exp\left(-\frac{2r^2}{W^2}\right)\exp\left\{2.22\sigma_1^2\left[\frac{2L}{kW^2}\right]^{\frac{5}{6}}\left[\frac{r^2}{W^2}\right] - 1.33\sigma_1^2\left[\frac{2L}{kW^2}\right]^{\frac{5}{6}}\right\}, \tag{2}$$

where the $W(L)$ is the radius of Gaussian spot and the W_0 is the waist radius. The k is wave number, and the σ_1^2 is the Rytov variance which has the following relationships with refractive index structure constant C_n^2

$$\sigma_1^2 = 1.23C_n^2 k^{7/6}L^{11/6} \tag{3}$$

Based on zero inner scale model, the $\sigma_1(r,L)$ has the following expression:

$$\sigma_1^2(r,L) = 4.42\sigma_1^2\Lambda_e^{5/6}\frac{r^2}{W_e^2} + \exp\left[\frac{0.49\sigma_B^2}{\left(1+0.56\sigma_B^{12/5}\right)^{7/6}} + \frac{0.51\sigma_B^2}{\left(1+0.69\sigma_B^{12/5}\right)^{7/6}}\right] - 1, \tag{4}$$

where the $W_e = W\left[1 + 1.63\sigma_1^{12/5}\cdot\left(2L/kW^2\right)\right]$ is the effective radius of Gaussian spot under turbulence, and σ_B^2 is the Rytov variance for Gaussian beam wave, which can be approximately defined by:

$$\sigma_B^2 = 3.86\sigma_1^2\left\{0.40\left[(1+2\Theta)^2 + 4\cdot\left(\frac{2L}{kW^2}\right)^2\right]^{5/12}\times\cos\left[\frac{5}{6}\tan^{-1}\left(\frac{1+2\Theta}{4L/kW^2}\right)\right] - \frac{11}{16}\left[\frac{2L}{kW^2}\right]^{5/6}\right\}. \tag{5}$$

The parameter Λ_e and Θ shown in the equation can be written as:

$$\Lambda_e = \frac{2L/\left(kW^2\right)}{1 + 1.63\sigma_1^{12/5}\cdot 2L/\left(kW^2\right)}, \tag{6}$$

$$\Theta = 1 + \frac{L}{R(L)}. \tag{7}$$

The $R(L) = L\left[1 + \left(\frac{\pi W_0^2}{L\lambda}\right)\right]$ is the radius of curvature. Using the above formula, the PDF of irradiance is obtained. According to Reference [23], values of C_n^2 near the ground in warm climates generally vary between 10^{-14} to 10^{-12} m$^{-2/3}$. So C_n^2 is selected as 10^{-13} m$^{-2/3}$. Therefore, the PDF of irradiance is shown as below, under the conditions where the transmitting angle is 1 mrad, the alignment error r = 0, the transmitting optical power is 1 W and the communication distance L = 1 km:

$$p(I) = \frac{1}{1.98 \times I}\exp\left[-\frac{(\ln I + 0.089)^2}{1.25}\right]. \tag{8}$$

The relationship between the probability distribution function of irradiance and irradiance is shown in Figure 2.

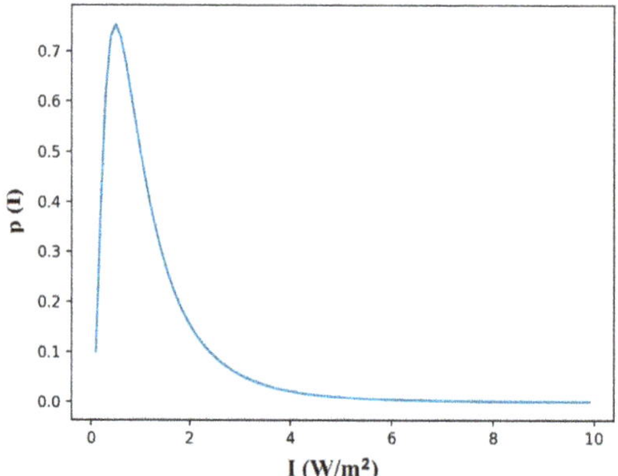

Figure 2. Probability distribution function of irradiance.

According to Reference [24], the coherence time of typical weak atmospheric turbulence is 1–10 ms, while the time of sending a single packet of data in the system is less than 1 us, so it can be concluded that there will be no sudden change in the time channel in the process of sending a packet of data. When the receiving area S = 1 cm^2 and the detection sensitivity w = 10 μW, the minimum irradiance that can be detected by the receiving end is

$$I_{min} = w/S = 0.1 \text{ W/m}^2. \tag{9}$$

So the communication probability of a single packet is

$$P = \int_{I_{min}}^{\infty} p(I)dI = 54.27\%. \tag{10}$$

The relationship between the number of times n that the system sends the same data packet in 1 s and the communication probability of the system in 1 s is as follows:

$$P_{t=1s} = 1 - (1 - P)^n = 1 - 0.4573^n. \tag{11}$$

The function image is shown in Figure 3.

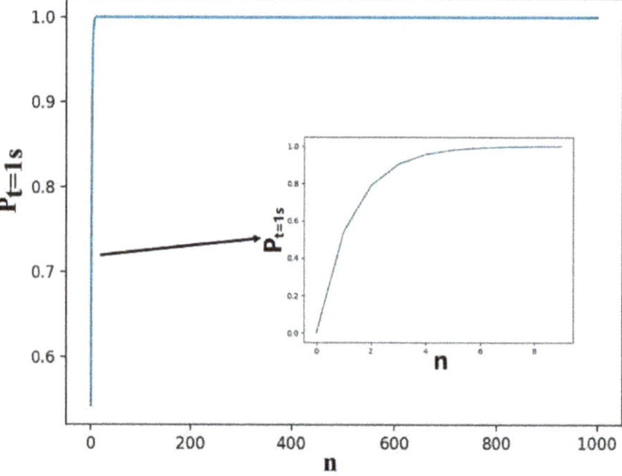

Figure 3. The relationship between the communication probability and the number of transmissions in 1 s.

The communication rate of the system is 1 Mbps, and the length of each packet is 1 Kb. So, the same packet can be sent 1000 times in 1 s. The communication probability of the system in 1 s is:

$$P_{t=1s} = 1 - P^{1000} = 99.99\%. \tag{12}$$

According to the calculation, although the UAV will be affected by the atmospheric turbulence in the flight process, the probability of the UAV receiving the command in 1 s is greater than 99.99%. The reason is that the coherent time is much longer than the sending time of single packet data, and the command and control data packet is sent repeatedly in a short time. Moreover, the communication performance of the system is good.

3. Design

The receiving sub modules are composed of short focus lenses, optical filters, detectors and amplification circuits, as shown in Figure 4. The signal light transmitted in space converges to the back of the filter after passing through the short-focus lens. By doing so, it can make the normal incident light spot larger in case of burning the filter and detector. What is more, the detector can receive more oblique incident light energy. It also enables the effective reception of optical signals in the range of 0–30° incident angle. The signal light is collected by the detector after passing through the filter. The detector converts the signal light into current signal. Then the receiving sub module outputs the signal to the core board circuit in the form of voltage signal after two-stage amplification circuit.

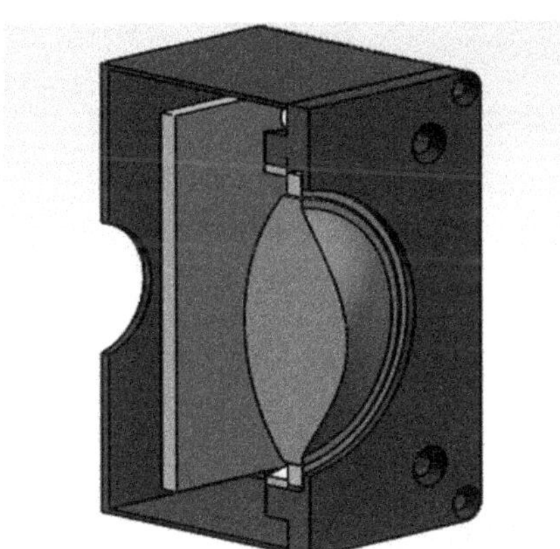

Figure 4. Receiving sub module model.

Since the window angle of a receiving sub module is only 30°, multiple receiving sub modules are needed to realize field of view splicing. According to the calculation, 12 receiving sub modules can realize $360° \times 30° \left(\theta_{\parallel} \times \theta_{\perp} \right)$ circular reception, and 52 receiving sub modules can realize real omnidirectional reception. These sub modules are fixed in different positions of the spherical support and work independently to achieve omnidirectional receiving function, as shown in Figure 5.

Each voltage signal is input to the core board circuit after the amplifying and the converting to digital signal by Analog-to-Digital Converter (ADC). Field Programmable Gate Array (FPGA) selects the strongest input signal while monitoring multiple signals in real time, and it is the command and control signal which is needed. If the signal strength of one channel exceeds the signal being collected in the detection process, and the strongest signal is similar to the signal being collected, the current strongest signal is used to replace the previous signal to make sure that the command and control signal being collected is the strongest one from all signals. The whole system continues monitoring the signal in the whole process.

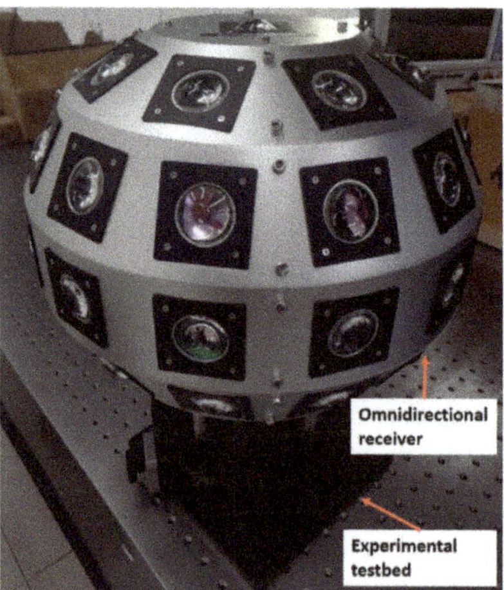

Figure 5. System model of omnidirectional reception.

According to the structural design of field of view splicing, if the whole system vibrates due to external influence, the incident signal light will move from one receiving sub module to another. In this process, the light received by the former module gradually becomes weak, and the light received by the latter module gradually becomes stronger, and the two are very similar. When the output signal strength of the latter exceeds that of the former, the signal collected by the whole system will also be switched to ensure that the collected command and control signal is the strongest effective one. Therefore, the influence of external vibration on the whole system can be compensated by the design of the system. The process of switching command and control signals in the system is realized by FPGA program design. During the process of FPGA, the pure static switching can be realized without generating the moment of inertia and interference to the flight process of UAV, which ensures the flight safety of UAV.

Because of the omnidirectional receiving design, the optical signal transmitted can be effectively received in any direction. Once the communication link being connected is interrupted, the receiver of omnidirectional receiving can receive the retransmitted link signal from the ground control center quickly, which can realize the fast link establishment. It will not occur when the receiver cannot receive the signal effectively for lacking of alignment at the receiver when the link is interrupted.

When the parallel light passes through the short focus lens, due to the influence of the non-uniform temperature distribution in the environment, the lens will undergo thermal deformation. It will cause the wave-front distortion of the parallel light, and result in the intensity and weakness of the focused spot [25]. Compared with the optical fiber or small area detector, the large area detector can receive a larger range of signal light. This kind of distorted signal has less influence on the large area detector, which makes the system have stronger anti-turbulence ability.

4. Experiment

A prototype was built to demonstrate the omnidirectional receiving ability. The light received by the receiving sub module converges on the filter through a short focal lens. When the incident light angle is moved, the relationship between the light intensity received by the detector and the departure angle is described as:

$$P_1 = P_0 \cos(\theta - 15°) \quad 0° \leq \theta \leq 30° \tag{13}$$

where P_0 is the intensity of the normal incident light, and θ is the angle between the incident light and the normal incident light when the incident light is moved. When the light moves, the spot received by the detector will first move from the edge to the center of the detector, and then gradually move out of the detector. Therefore, the ratio of the spot area on the detector to the whole spot area S/S_0 will first increase, then remain unchanged, and finally decrease. The light intensity received by the detector is:

$$P = P_1 \times \frac{S}{S_0} \tag{14}$$

The output electrical signal of the module is proportional to the intensity of the incident light. When the incident light is moved, the relationship between the output voltage of the receiving sub module and the departure angle can be denoted as:

$$U = U_0 \cos(\theta - 15°) \times \frac{S}{S_0} \quad 0° \leq \theta \leq 30° \tag{15}$$

where U_0 is the output voltage at normal incidence. In Figure 6, the relationship between the departure angle and the output voltage theoretically is represented by the blue line. At the same time, the output voltage of a receiving sub module at different departure angles from 0° to 30° is actually tested as shown by the orange line in Figure 6. It can be seen from Figure 6 that when the signal light is obliquely incident at 0° to 8° and 22° to 30°, the theoretical data are in good agreement with the measured results. When the signal light is approximately normal incidence at 8° to 22° the theoretical data have a little deviation from the measured results, and the closer the signal light is to normal incidence, the greater the deviation is. This is because the filter will affect the oblique incident signal light. With the increase of the incident angle, the central wavelength and passband of the filter will move to the short wave direction [26]. Therefore, in order to ensure the signal quality of oblique incidence, the window wavelength of the filter of the receiving sub module is slightly biased to the short wave direction. In this way, the wavelength of the signal light and the window wavelength of the filter cannot be completely matched in normal incidence. This will make the filter have some attenuation to the signal and make the measured result lower than the signal strength of the theoretical data. However, because the light intensity of normal incidence is stronger than that of oblique incidence, the receiving sub module can still receive good signal even the filter has attenuation effect.

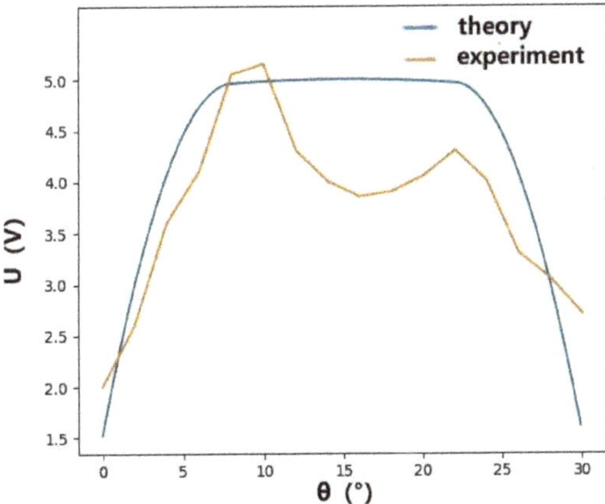

Figure 6. Relationship between the output voltage of the receiving sub module and the departure angle.

Then, the loss in the switching process of the combined optical system is analyzed. Since the window angle of each receiving sub module is 30°, 12 modules are used to achieve $360° \times 30° \left(\theta_\parallel \times \theta_\perp \right)$ circular reception. Consequently, when the direction of the incident light moves on the annular band, the relationship between the output voltage and the offset angle of the combined optical system changes periodically with a period of 30°. The image of each cycle is the same as that of a single receiving sub module.

In Figure 7, theoretically, the relationship between the departure angle and the output voltage is represented by the blue line. At the same time, the output voltage of the system at different departure angles from 0° to 180° is tested as shown by the orange line in Figure 7. The inconsistent performance of each receiving sub module in Figure 7 is caused by manual errors in circuit welding, packaging, etc. Although there are errors in the experiment, when the departure angle changes continuously, the prototype can continuously output the signal. It shows that although loss exists in the switching process, it does not have a serious impact on the signal reception and output of the whole model.

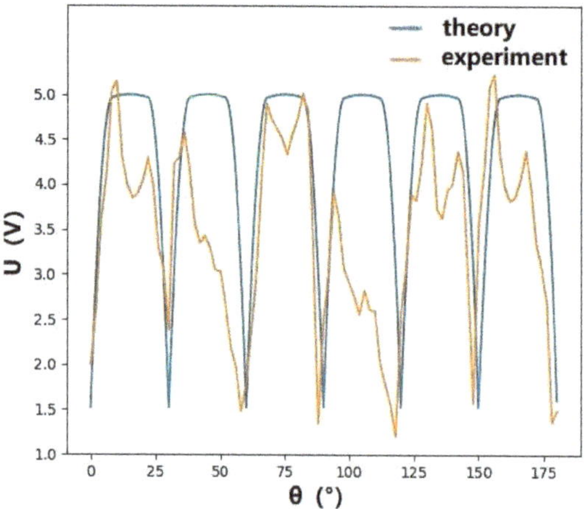

Figure 7. Relationship between the output voltage of the combined optical system and the departure angle.

Compared with the directional receiver in Reference [27], the system can realize omnidirectional reception. The system can be miniaturized without the acquisition-tracking-pointing, the erbium-doped fiber amplifier or the beaconing light system, and the system has stronger anti-turbulence capability than the directional receiver. However, due to the limited performance of large area detector, its communication rate is far lower than that of the directional receiver. Therefore, the system is suitable for some occasions where there is no need for high communication rate, but it needs to be miniaturized and omnidirectional.

5. Conclusions

An electromagnetic immune UAV command and control system based on FSOC is established. The system adopts the scheme of omnidirectional receiving. It has strong anti-turbulence ability with large area detectors and short focus lenses. The design of omnidirectional communication improves the ability of anti-vibration and link establishment. Pure static receiving has no momentum effect on the platform. The receiver is miniaturized without the gimbal mirror system, the beacon camera system, the QPD or the multi-level lens system. This work enables the FSOC system to be applied to UAV command and control, satellite submarine communication and other occasions requiring small FSOC equipment.

Author Contributions: Conceptualization, Y.Z. and Y.W.; hardware, Y.Z.; software, Y.W.; validation, Y.Z., Y.W., Y.D. and A.D.; writing—original draft preparation, Y.Z.; writing—review and editing, Y.Z. and Y.W.; project administration, J.L. All authors have read and agreed to the published version of the manuscript.

Funding: This research was funded by the National Key R&D Program of China under Grant 2019YFB2203700, and the National Nature Science Fund of China under Grants 61527820 and 61625504.

Data Availability Statement: Data is contained within the article.

Acknowledgments: The author thank Wenqi Yu for his fruitful discussions.

Conflicts of Interest: The funders had no role in the design of the study; in the collection, analyses, or interpretation of data; in the writing of the manuscript, or in the decision to publish the results. The authors declare no conflict of interest.

References

1. Alzenad, M.; Shakir, M.Z.; Yanikomeroglu, H.; Alouini, M.S. FSO-based vertical backhaul/fronthaul framework for 5G+ wireless networks. *IEEE Commun. Mag.* **2018**, *56*, 218–224. [CrossRef]
2. Hanzo, L.; Haas, H.; Imre, S.; O'Brien, D.; Rupp, M.; Gyongyosi, L. Wireless Myths, Realities, and Futures: From 3G/4G to Optical and Quantum Wireless. *Proc. IEEE* **2012**, *100*, 1853–1888. [CrossRef]
3. Haas, H.; Yin, L.; Wang, Y.; Chen, C. What is LiFi? *J. Lightwave Technol.* **2015**, *34*, 1533–1544. [CrossRef]
4. Ding, J.; Chih-Lin, I.; Xu, Z. Indoor Optical Wireless Channel Characteristics with Distinct Source Radiation Patterns. *IEEE Photonics J.* **2016**, *8*, 1–15. [CrossRef]
5. Ding, J.; Chih-Lin, I.; Zhang, C.; Yu, B.; Lai, H. Evaluation of Outdoor Visible Light Communications Links Using Actual LED Street Luminaries. In Proceedings of the 13th Chinese Conference on Biometric Recognition, Urumqi, China, 11–12 August 2018.
6. Tosun, A.; Nouri, H.; Uysal, M. Experimental Investigation of Pointing Errors on Drone-based FSO Systems. In Proceedings of the 2020 28th Signal Processing and Communications Applications Conference (SIU), Gaziantep, Turkey, 5–7 October 2020.
7. Li, Z.; Wei, H. Study on interruption probability of FSOC in atmospheric turbulence inclined path channel. *Mod. Electron. Tech.* **2020**, *43*, 6–9.
8. Yang, Y.; Chen, M.; Guo, C.; Feng, C.; Saad, W. Power efficient visible light communication (VLC) with unmanned aerial vehicles (UAVs). *IEEE Commun. Lett.* **2019**, *23*, 1272–1275. [CrossRef]
9. Dong, Y.; Hassan, M.Z.; Cheng, J.; Hossain, M.J.; Leung, V.C. An edge computing empowered radio access network with UAV-mounted FSO fronthaul and backhaul: Key challenges and approaches. *IEEE Wirel. Commun.* **2018**, *25*, 154–160. [CrossRef]
10. Yokota, N.; Yasaka, H.; Sugiyasu, K.; Takahashi, H. Motion tolerance for dynamic object recognition using visible light IDs. In Proceedings of the 2018 IEEE 7th Global Conference on Consumer Electronics (GCCE), Nara, Japan, 9–12 October 2018; pp. 702–703.
11. Wu, D.; Sun, X.; Ansari, N. An FSO-based drone assisted mobile access network for emergency communications. *IEEE Trans. Netw. Sci. Eng.* **2019**, *7*, 1597–1606. [CrossRef]
12. Dautov, K.; Kalikulov, N.; Kizilirmak, R.C. The impact of various weather conditions on vertical FSO links. In Proceedings of the 2017 IEEE 11th International Conference on Application of Information and Communication Technologies (AICT), Moscow, Russia, 20–22 September 2017.
13. Cruz, P.J.; Fierro, R. Towards optical wireless communications between micro unmanned aerial and ground systems. In Proceedings of the International Conference on Unmanned Aircraft Systems, Denver, CO, USA, 9–12 June 2015.
14. Amantayeva, A.; Yerzhanova, M.; Kizilirmak, R. UAV location optimization for UAV-to-vehic le multiple access channel with visible light Communication. In Proceedings of the 2019 Wireless Days (WD), Manchester, UK, 24–26 April 2019; pp. 1–4.
15. Ukida, H.; Miwa, M.; Tanimoto, Y.; Sano, T.; Yamamoto, H. Visual UAV control system using LED panel and onboard camera. In Proceedings of the 2013 IEEE Instrumentation & Measurement Technology Conference, Minneapolis, MN, USA, 6–9 May 2013; pp. 1386–1391.
16. Li, L.; Zhang, R.; Liao, P.; Cao, Y.; Song, H.; Zhao, Y.; Du, J.; Zhao, Z.; Liu, C.; Pang, K.; et al. MIMO equalization to mitigate turbulence in a 2-Channel 40-Gbit/s QPSK free-space optical 100-m round-trip orbital-angular-momentum-multiplexed link between a ground station and a retro-reflecting UAV. In Proceedings of the 2018 European Conference on Optical Communication (ECOC), Rome, Italy, 23–27 September 2018; pp. 1–3.
17. An, J.; He, X.; Yang, Q.; Xu, L.; Liu, X. Research on the application of the air to ground free space optical communication by small UAV. *Opt. Commun. Technol.* **2017**, *41*, 10–13.
18. Liu, G.; Tao, D. Research on lateral oscillation of small UAV in high speed flight. *Electron. Opt. Control* **2021**, *28*, 97–99.
19. Toyoshima, M.; Takenaka, H.; Koyama, Y.; Takayama, Y.; Kunimori, H.; Kubooka, T.; Suzuki, K.; Yamamoto, S.; Taira, S.; Tsuji, H.; et al. Terrestrial free-space optical communications network for future airborne and satellite-based optical communications projects. In Proceedings of the 31st AIAA International Communications Satellite Systems Conference, Florence, Italy, 14–17 October 2013; p. 5686.

20. Anscombe, N. Adapting to change. *Nat. Photonics* **2011**, *5*, 18–19. [CrossRef]
21. Li, M.; Lin, S.; Li, S.M.; Yang, S.W. Suppression of fluctuation on the angle of arrival for free- space optical communication. *Optoelectron. Lett.* **2012**, *8*, 297. [CrossRef]
22. Andrews, L.C.; Phillips, R.L.; Hopen, C.Y. *Laser Beam Scintillation with Applications*; SPIE Press: Bellingham, WA, USA, 2001.
23. Michailidis, E.T.; Nomikos, N.; Bithas, P.; Vouyioukas, D.; Kanatas, A.G. Outage probability of triple-hop mixed RF/FSO/RF stratospheric communication systems. In Proceedings of the International Conference on Advances in Satellite & Space Communications, Athens, Greece, 22–26 April 2018.
24. Chen, J.; Ai, Y.; Tan, Y. Improved free space optical communications performance by using time diversity. *Chin. Opt. Lett.* **2008**, *6*, 797–799. [CrossRef]
25. Tong, L. Study on the Restoration Algorithm of Distorted Sports in Laser Link Between Satellites. Master's Thesis, Harbin Institute of Technology, Harbin, China, 2018.
26. Zhang, J.; Wang, X.; Zhang, X.; Yan, C.; Zhang, B. Changchun Institute of Optics, Fine Mechanics and Physics; Chinese Academy of Sciences. Effects and Corrections of Incidence Light Cone on Transmission Characteristics of Narrowband Filter. *Acta Opt. Sin.* **2014**, *34*, 307–311.
27. Liu, D.; Wang, Z.; Liu, J.; Tan, J.; Yu, L.; Mei, H.; Zhou, Y.; Zhu, N. Performance analysis of 1-km free-space optical communication system over real atmospheric turbulence channels. *Opt. Eng.* **2017**, *56*, 1. [CrossRef]

 *photonics*

Article

A Novel Data-Aided Frame Synchronization Method Based on Hough Transform for Optical Communications

Huiwen Yin, **Sida Li, Zhiping Huang * and Jie Chen**

College of Intelligent Science and Technology, National University of Defense Technology, Changsha 410073, China; yinhuiwen18@163.com (H.Y.); lisida_0301@163.com (S.L.); chenjie_nudt@126.com (J.C.)
* Correspondence: huangzhiping65@nudt.edu.cn; Tel.: +86-137-0731-3262

Received: 3 August 2020; Accepted: 24 August 2020; Published: 27 August 2020

Abstract: In optical communication systems, frame synchronization is essential for subsequent operations, such as error correction and payload extraction. Various methods, so far, have been proposed in the published literature, but the performance is unsatisfactory under high bit error rate (BER) conditions. We present, in this work, a novel data-aided frame synchronization technique for optical packet transmission systems, in which the transmitter sends a sequence of packets with a specific synchronization word periodically inserted, and the receiver blindly recognizes the synchronization word to attain frame synchronization. The proposed algorithm detects the synchronization word based on Hough transform (HT), a classic method for line detection in digital image processing. The core principle of the algorithm is to exploit the periodicity of the frame synchronization word, which appears as black-and-white spaced stripes on a binary image when the frames are all aligned. Simulations are conducted over a 56Gbps optical QPSK transmission system, and the results show that our algorithm is still effective in attaining frame synchronization at a BER of 0.3. A comparison is also made between our algorithm and existed autocorrelation (AC)-based algorithm. The results demonstrate that our algorithm has a better error resilience performance. When the BER is higher than 0.03, our algorithm outperforms an AC-based algorithm significantly.

Keywords: optical communication; frame transmission; synchronization word; blind recognition; Hough transform

1. Introduction

In optical communication systems, frame synchronization is indispensable to subsequent signal processing, such as error correction and information extraction [1,2]. It can be considered as a type of reverse engineering problem in the field of information theory, which has received increased attention over the past few decades [3]. There are mainly two categories of frame synchronization techniques: data-aided frame synchronization and code-aided frame synchronization. For the former, synchronization is attained through recognizing the synchronization word, while for the latter, synchronization is achieved through exploiting the properties of error correction code.

In cooperative communications, data-aided frame synchronization is commonly applied, and there are two main types of methods, correlation rule and maximum-likelihood (ML) rule [4]. In 1953, Barker first proposed a correlation-based recognizer to detect the given frame synchronization sequence in received bit stream [5]. This method is simple to implement, but it has a poor performance and a narrow application scope. Some improved schemes have also been put forward in succession, such as double correlation [6,7], which is robust to carrier frequency and phase errors. A method based on the ML rule for frame synchronization was first proposed by Massey in 1972 over additive white Gaussian noise (AWGN) channel [8], and he concluded that ML-based detection has a better performance than traditional correlation detection. The conclusion was subsequently confirmed in the environment of high SNR by Nielsen [9], and was further extended to the M-ary phase shift keying modulation by Liu and Tan [10]. Later, ML-based frame synchronization algorithms for flat fading channels and frequency-selective channels were proposed successively [11–13]. In addition to the correlation and ML rule, some other schemes have also been offered in recent years. Chiani and Martini applied the likelihood ratio test (LRT) in the hypothesis test theory to frame synchronization detection [14–16], and their method performs better than correlation in AWGN channels. Kim exploited post-detection integration (PDI) to provide coarse frame alignment with limited complexity with respect to ML solutions [17]. More recently, another novel algorithm called balanced generalized PDI (B-GPDI), which combines the ML rule and PDI for synchronization detection, was reported in [18].

In non-cooperative communications, a synchronization word needs to be recognized blindly for data-aided frame synchronization. Yang proposed a blind recognition algorithm based on the peak characteristic of the partial third-order correlation function for the synchronization word being m-sequence [19]. A joint recognition algorithm based on correlation filtering and Hadamard transform is proposed in [20], and it performs well with a high bit error rate. Naseri used the auto-correlation function to blindly recover the synchronization word and frame length in a satellite system, and proposed another two algorithms based on hardware equipment, in case auto-correlation cannot determine the exact frame length [21]. The blind recognition algorithm of the synchronization word based on soft-decision for QPSK signals was proposed in [22], and it can clearly enhance the accuracy of blind recognition. Guo proposed a cumulative filtering algorithm to identify the frame header [23], but it is difficult to determine the decision threshold with a complex frame structure or uncertain synchronization word length. To resolve this issue, an improved algorithm based on autocorrelation (AC) was proposed by Xu [24]. Peak-to-average ratio (PAR) is calculated in his algorithm to overcome high bit-error-rate (BER). In recent years, code-aided frame synchronization has received considerable attention, due to the advantages of saving bandwidth and high spectral efficiency [25–28].

In this paper, we propose a novel frame synchronization method based on Hough transform, in which synchronization is achieved through blindly recognizing the synchronization word. In our work, we build an interception matrix filled with intercepted bit stream. When frames are all aligned in the interception matrix, the columns of synchronization word appear as black and white stripes, if we consider the matrix as a binary image. Hence, Hough transform, which is a classic method for line detection in digital image processing, is applied in this work to detect the stripes in the interception matrix. Simulations are carried out in the optical QPSK transmission system, and the results show that the new method can efficiently recognize the synchronization word and attain frame synchronization. The proposed method is also compared with the AC-based algorithm and simulation results demonstrate that our HT-based algorithm performs better at high BER.

The remainder of this paper is organized as follows. Section 2 briefly introduces the existing AC-based frame synchronization algorithm. In Section 3, principles of Hough transform are first introduced in detail and an example of its application is given. Then, the data-aided frame synchronization algorithm based on Hough transform is presented. Simulation experiments are carried out in Section 4, and the HT-based synchronization algorithm is compared with the AC-based algorithm. Finally, Section 5 provides the conclusion.

2. Fundamental of AC-Based Frame Synchronization Algorithm

Information is usually transmitted using frames as basic unit in the optical transmission systems. Figure 1a gives a schematic diagram of frame sequence with synchronization word inserted centrally. It can be shown that an intact frame, the length of which is L, contains two parts, fixed frame synchronization word (S) and varying payload (D). The length of them are M and N, respectively, and $L = M + N$. In the data received by receiver, the synchronization word is periodic emergence, while the payload is random. Through locating the synchronization word, frame synchronization is attained, and the payload can be extracted further for subsequent operations.

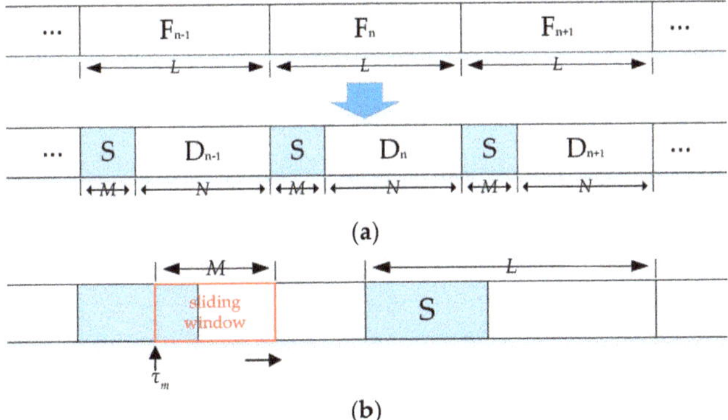

Figure 1. Received frame sequence. F, frame; S, frame synchronization word; D, payload; *L*, frame length; *M*, synchronization word length; *N*, payload length; (**a**) diagram of frame sequence with fixed synchronization word inserted.; (**b**) sliding window starting from τ_m.

In cooperative communications, S is known to the receiver and synchronization is achieved through finding the optimal delay that maximizes the cross-correlation of received sequence and S. The high SNR approximation of this method is given in [8]:

$$\hat{\tau} = \underset{\tau_m}{\arg\max}\frac{1}{M}\left|\sum_{i=0}^{M-1} r_{\tau_m}^i s_i^*\right|^2 \tag{1}$$

where s_i is the *i*th element of S and $r_{\tau_m}^i$ is the *i*th element of received sequence in the sliding window, starting from τ_m, as shown in Figure 1b.

However, in the context of non-cooperative communications, S is unknown to the receiver. Hence, the cross-correlation in Equation (1) is not computable. An alternative option using auto-correlation is given in [24]:

$$\left(\hat{\tau}, \hat{L}\right) = \underset{\tau_m,\, l}{\arg\max}\frac{1}{l}\left|\sum_{i=0}^{l-1} r_{\tau_m}^i \left(r_{\tau_m+l}^i\right)^*\right|^2 \tag{2}$$

where $r_{\tau_m}^i$ and $r_{\tau_m+l}^i$ are the *i*th element of received sequence in the sliding window starting from τ_m and τ_m+l, respectively, l is the length of sliding window and $\hat{L}$ is an estimate of the frame length.

Specifically, a sampling data storage matrix is built in the AC-based synchronization algorithm, and $r_{\tau_m}^i r_{\tau_m+l}^i$ is sequentially filled into the matrix. Since the probability distributions of 0 and 1 in S and D are different, the synchronization word can be detected through calculating the probability distributions of 0 and 1 at each position of the sliding window, with τ_m and the l constant.

3. HT-Based Frame Synchronization Algorithm

3.1. Hough Transform

In 1962, Hough devised an ingenious method, popularly called Hough transform (HT), which efficiently identifies lines in images [29,30]. It is still an important tool for digital image processing nowadays. The core idea of Hough transform is to establish a dual relationship between points and lines, image space and Hough space. The principle of detecting straight lines based on Hough transform is briefly introduced as follows.

The equation representation of a straight line in cartesian coordinates (we call it *xy*-space in this paper) can be expressed by slope (*k*) and intercept (*b*) as follows.

$$y = kx + b \tag{3}$$

If the parameter space, *kb*-space is established, the straight line in *xy*-space actually corresponds to a point in *kb*-space. Similarly, a straight line in *kb*-space corresponds to a point in *xy*-space. This is a so-called dual relationship mentioned above. It can be easily drawn that intersecting position relationship in *kb*-space corresponds to collinear position relationship in *xy*-space. However, there is a parameter problem with the established *kb*-space. The vertical lines in *xy*-space, whose slope is infinite, do not have corresponding points in *kb*-space. To this end, space is established and the correspondence can be written as.

$$\rho = x \cos\theta + y \sin\theta \tag{4}$$

As shown in Figure 2, ρ is the algebraic distance from the origin to the line and θ is the angle between *x*-axis and the vector perpendicular to the line. If we restrict θ to the interval $(-90°, 90°)$, the dual relationship between *xy*-space and $\rho\theta$-space exists in any cases. However, the curve in $\rho\theta$-space corresponding to the point in *xy*-space is not a straight line, but a sinusoid.

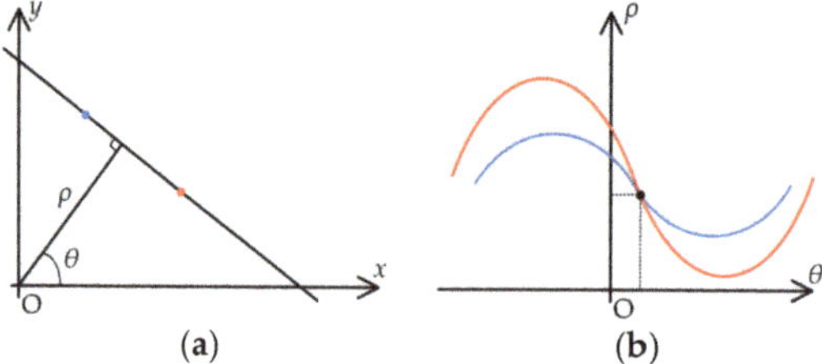

Figure 2. Dual relationship between *xy*-space and $\rho\theta$-space. (**a**) *xy*-space; (**b**) $\rho\theta$-space.

The *xy*-space is the image space while the $\rho\theta$-space is sometimes called Hough space. The properties of point-to-line dual relationship between image space and Hough space can be summarized as follows.

- A point in image space corresponds to a sinusoid in Hough space;
- A point in Hough space corresponds a straight line in image space;
- The collinear position relationship of points in image space corresponds to the intersecting position relationship of curves in Hough space.

However, it is essentially exhaustive that we map all the points in image space into their corresponding curves in Hough space, without any restrictions for straight line detection, and the computation required grows quadratically with the number of points [31]. To this end, Hough space is quantized into grids, and accumulator cells are created. For each non-background point (x_n, y_n) in image space, the corresponding curve given by Equation (4) is entered in the grids by incrementing the count in each accumulator cell along the curve. Eventually, a given cell records the total number of curves passing through it, which means that these points in image space are collinear. Figure 3 gives an example for a thorough understanding.

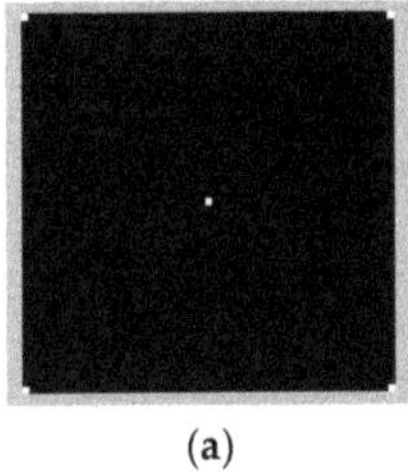

(a)

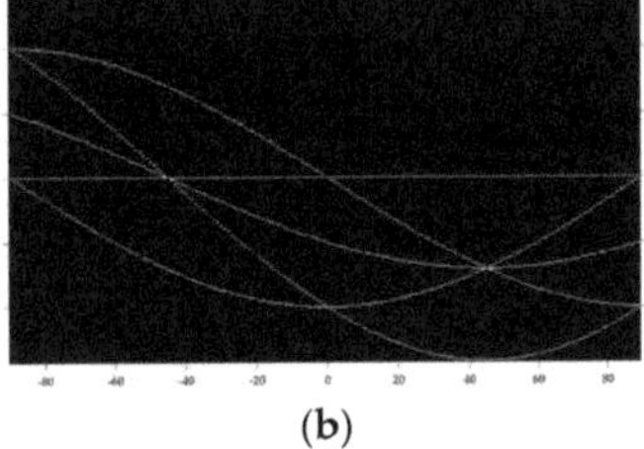

(b)

Figure 3. Example of line detection based on Hough transform (HT). (**a**) Image space with five points in a black background; (**b**) Hough space.

There are five points in the image space with a black background, which corresponds to five curves in Hough space, respectively. θ is quantified at 1-degree intervals and the interval for ρ is also one. Hough space is divided into accumulator cells. It can be seen from Hough space that the two points at $\theta = \pm 45°$ are generated by the intersection of the three curves, which means that there are two sets of three collinear points in the image space. The four sets of collinear points on the vertical and horizontal lines in image space corresponds to four points generated by the intersection of two curves in Hough space.

3.2. Algorithm

In the context of non-cooperative communications, M, N, and the type of S are all unknown, which require to be recognized blindly for frame synchronization [32]. Blind recognition of frame parameters can be implemented through the properties of frame synchronization word, which are fixed and cyclical. To achieve this, the interception matrix C, with r rows and l columns, is built by filling it with the intercepted bit stream from top left to bottom right. As shown in Equation (5), Y is the intercepted bit stream.

$$Y = \begin{pmatrix} y_1, & y_2, & y_3, & \cdots \end{pmatrix} \overset{fill\ in}{\Rightarrow} \begin{pmatrix} y_1 & y_2 & \cdots & y_l \\ y_{l+1} & y_{l+2} & \cdots & y_{2l} \\ \vdots & \vdots & \ddots & \vdots \\ y_{(r-1)l+1} & y_{(r-1)l+2} & \cdots & y_{rl} \end{pmatrix} = \begin{pmatrix} c_{1,1} & c_{1,2} & \cdots & c_{1,l} \\ c_{2,1} & c_{2,2} & \cdots & c_{2,l} \\ \vdots & \vdots & \ddots & \vdots \\ c_{r,1} & c_{r,2} & \cdots & c_{r,l} \end{pmatrix} = C \quad (5)$$

There are two cases for the number of columns of interception matrix C, $l \neq kL$ ($k \in Z^+$) and $l = kL$. In case of $l = kL$, all elements of certain columns are either 0 or 1, which are called synchronization-word columns in this paper, while elements of other columns are random, and we call these non-synchronization-word columns. In the case of $l \neq kL$, there do not exist synchronization-word columns yet. Based on this property, all we need to do is to go through all possible values of l to estimate the frame length L and synchronization word position.

As we all know, $y_i = 0$ or 1 ($i \in Z^+$), which means the interception matrix C is filled with 0 and 1. If we consider C as the image space, in which 0 is black and 1 is white, it can be displayed as a binary image in Figure 4. The two cases, that $l \neq kL$ and $l = kL$, are shown in Figure 4a,b, respectively. The synchronization-word columns appear as black and white straight stripes in a binary image, while the non-synchronization-word columns appear as snowflake noise.

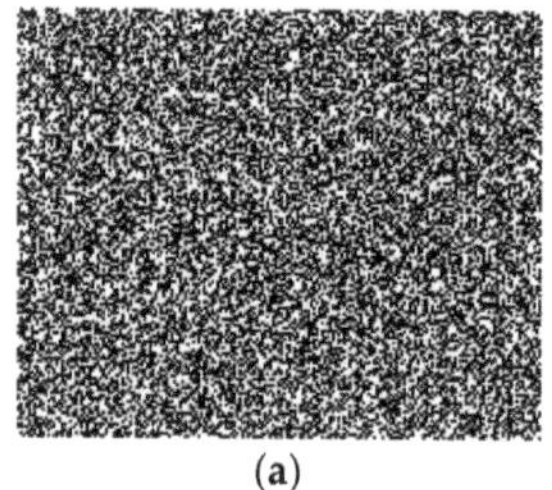 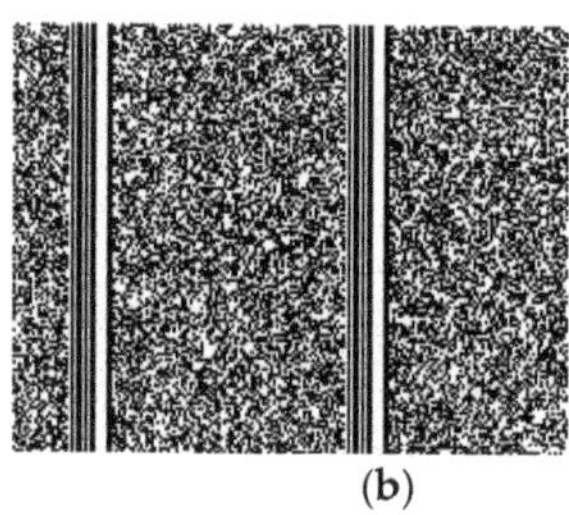

(a)   (b)

Figure 4. Binary image of interception matrix C. (a) $l \neq kL$; (b) $l = kL$.

To identify the case that $l = kL$ when going through all possible values of l, line detection based on Hough transform can be applied. However, there practically exist collinear points at any angle in the image space C, which means straight lines can be detected with taking any values. Considering that the number of collinear points in the synchronization-word columns is the largest in the vertical direction, we just need to simply find the accumulator cells with the most counts at $\theta = 0°$ in Hough space for all l. The recognition algorithm is summarized as follows.

Step1. Set the search range of l and initialize it. (Assume that $l \in [l_{min}, l_{max}]$);

Step2. Fill the image space $C_{r \times l}$ with the intercepted bit stream Y from top left to bottom right;

Step3. Perform edge extraction on image space;

Step4. Quantify θ and ρ. For each non-background point in image space $C_{r \times l}$, calculate its corresponding sinusoid according to Equation (4) and fill the curve into Hough space by incrementing the count in each accumulator cell along the curve;

Step5. Record the largest counts among all the accumulator cells at $\theta = 0°$ in Hough space. Let it be x_l.

Step6. If $l < l_{max}$, let $l = l + 1$ and return to Step2. Otherwise, execute Step 6;

Step7. Fill all the x_l into vector X and find the maximum value. Let it be x_{max};

$$X = \begin{pmatrix} x_{l_{min}}, & x_{l_{min}+1}, & \cdots, & x_{l_{max}} \end{pmatrix} \tag{6}$$

Step8. Extract the elements greater than αx_{max} from the vector X, and find the one with the smallest subscript from them, and use the subscript value as the estimate of the frame length L, which is recorded as $\hat{L}$. (α will be determined in the simulation next);

Step9. Let $l = \hat{L}$ and fill the image matrix $C_{r \times l}$ with the intercepted bit stream Y;

Step10. Record the count of each accumulator cell at $\theta = 0°$ in Hough space and recognize the synchronization word.

4. Simulation Results and Discussion

In this section, the performance of HT-based data-aided frame synchronization method is analyzed. Simulations are conducted over a 56 Gbps QPSK coherent optical transmission system built in Optisystem. The BER of received bit stream is first calculated at the receiver, and then the frame synchronization algorithm is applied to attain synchronization before payload extraction.

The frame length L in simulation is set to 1024 bits, while the frame synchronization word S is 0xA53C, 16 bits in length. The payload D is generated randomly. Details of how the proposed algorithm works are shown first, and then its performance in different situations is analyzed. Finally, a comparison between our proposed algorithm and the algorithm proposed in [24] is made.

First, the workflow of HT-based blind recognition algorithm is presented at BER = 2×10^{-2}. r, the rows of image space, is set to 100 and the search range of l is [500,5000]. Figure 5 above gives the running results of Step2–Step4 of the algorithm at $l = 1024$. It can be seen from the Step 2-result that there is obvious image noise in the synchronization-word columns, which is circled with a red frame. From the Step3-result, we can find that the noise is eliminated after edge extraction, and the characteristics of the synchronization-word columns are retained. Apparently, our algorithm can suppress the effects of noise to some extent, thus improving the error resilience performance.

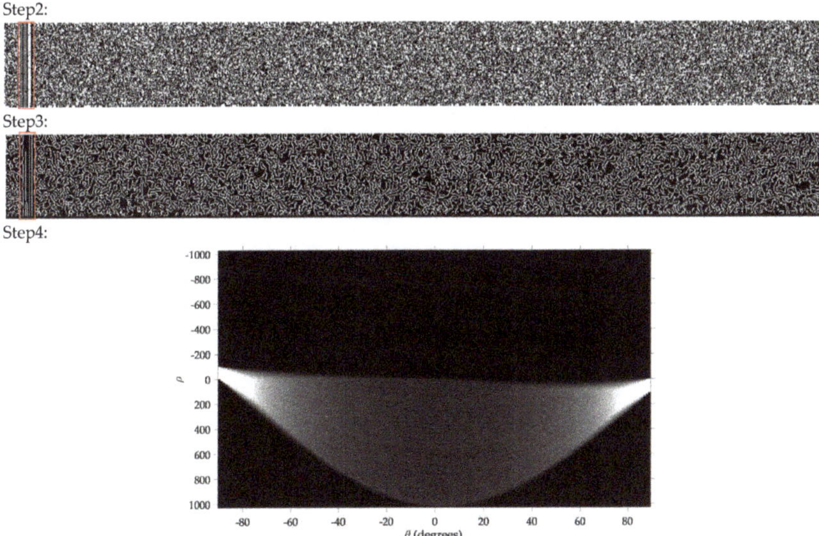

Figure 5. Running results of Step2–Step4. Result of Step2 is the binary image of image matrix C and the result of Step3 is the binary image after edge extraction; the result of Step4 is the Hough space.

Figure 6 gives the running result of frame length search in Step7. It can be seen that the case that $l = kL$ ($k = 1,2,3,4$) is effectively identified. Considering that the one with the smallest subscript among those greater than αx_{max} is taken as $\hat{L}$ in Step8, α should be greater than $\frac{75}{98}$ (0.77) so that $\hat{L} = L$. Result of synchronization-word recognition in Step10 is shown in Figure 7. The synchronization-word columns are recognized and since 0 is background point, it can be seen from the figure that the synchronization word is 1010,0101,0011,1100 (0xA53C). From Figures 5–7, we can find that HT-based blind recognition algorithm can effectively recognize the frame parameters, including frame length L, frame synchronization word S and its length M, under the BER condition of the error correction limit (BER = 2×10^{-2}).

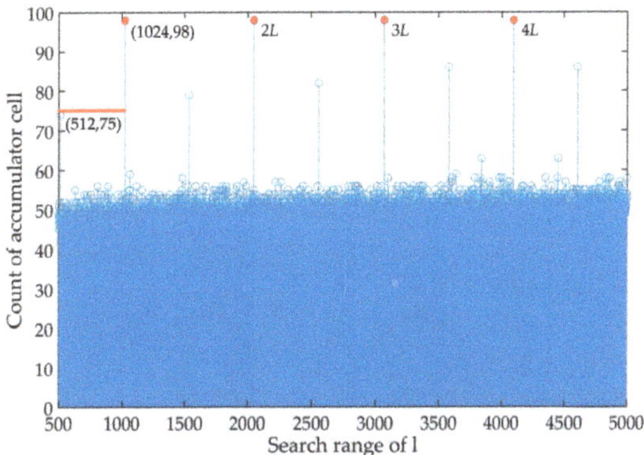

Figure 6. Result of frame length search in Step7. ($l \in$[500,5000] and L = 1024)

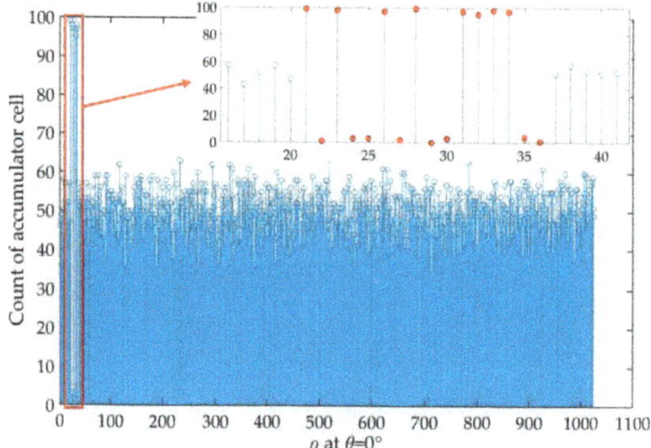

Figure 7. Result of frame synchronization word recognition in Step10. The enlarged part is the synchronization-word columns.

Then, the value of α is discussed at different BER, and we have also explored the influence of the selection of r on the value of α. For each BER level, 20 simulations are conducted, and α takes the mean. The curves of α versus BER for different values of r are shown in Figure 8. When r takes 20, the value of α is greater than 1, as BER exceeds 0.2, which means that the frame parameters cannot be recognized correctly. When r exceeds 50, our algorithm still works at BER = 0.3. On the other hand, when r exceeds 200, the value of r has little influence on α. Hence, the optimal value of r should be between 100 and 200, taking into account both the amount of calculation and the accuracy of identification. In addition, it can be seen from Figure 8 that the value of α decreases first, and then turns to increase as BER increases, which reflects the ability of our algorithm to suppress noise from another aspect. When the value of r increases, the turning point of α moves towards higher BER.

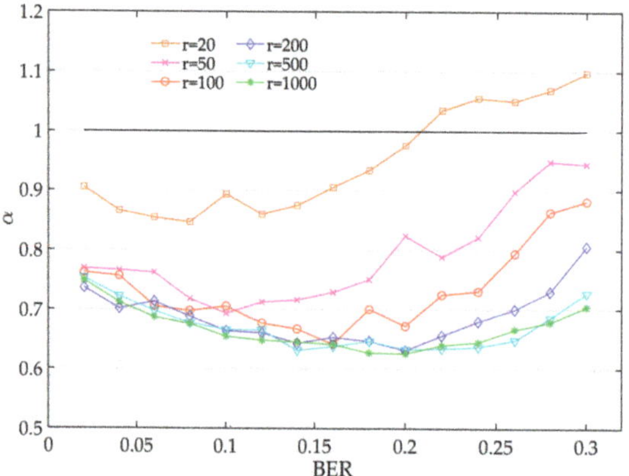

Figure 8. Value of α versus bit error rate (BER) under different r. r, the columns of interception matrix C.

Finally, the recognition accuracy, which is defined as the ratio of the times of recognizing correctly to the total times, is compared between our proposed algorithm and the autocorrelation-based recognition algorithm proposed in [24]. A total of 100 simulations are carried out for each BER level. The value of α takes 0.8 and r is set to 100 in the comparison simulation, according to Figure 8. The results are shown in Figure 9.

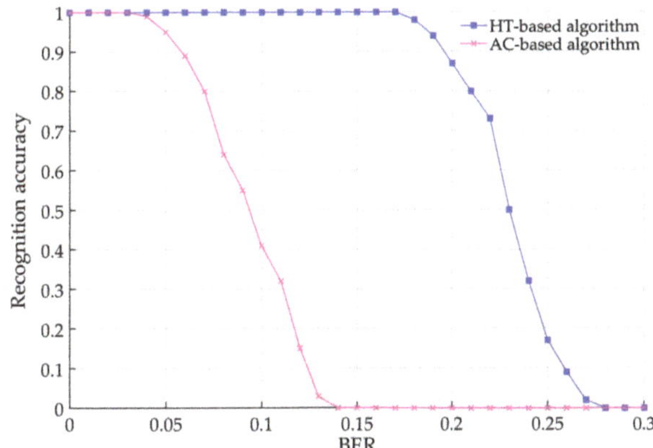

Figure 9. Comparison between HT-based recognition algorithm and AC-based algorithm. HT, Hough Transform; AC, autocorrelation.

As shown in Figure 9, the HT-based blind recognition algorithm outperforms the autocorrelation-based blind algorithm. When the BER is over 0.1, the recognition accuracy of the AC-based algorithm is lower than 0.5, while that of the HT-based algorithm is still 1. The recognition accuracy of HT-based algorithm starts to gradually decrease at BER = 0.18 and deteriorates sharply at BER = 0.22. Actually, the recognition accuracy is related to the value of α. Apparently, our algorithm has a better error resilience performance.

5. Conclusions

In this work, we have conducted a detailed investigation on the techniques of frame synchronization, and give a brief review of some previous works. A typical frame synchronization method based on auto-correlation was introduced as a comparison. Our proposed algorithm based on Hough transform was finally presented. Simulations were carried out over a 56Gbps QPSK coherent optical transmission system.

The workflow of our algorithm was shown first and some key parameters were analyzed, including r, the columns of interception matrix C, and α, a threshold set for recognizing frame length. The results show that our algorithm still works well at BER = 0.3, and the optimal value for r is between 100 and 200, considering both the computation and bit-error resilience. The trend of the value of α with the growth of BER reflects the suppression effect on noise of our algorithm. A comparison of recognition performance between our algorithm and an existing autocorrelation-based algorithm was also made. The results demonstrate that our algorithm outperforms the AC-based algorithm significantly, as the BER exceeds 0.03. It is noteworthy that our algorithm is also effective for the recognition of the frame synchronization word inserted in the frame at intervals.

However, the proposed algorithm requires the frame synchronization word to be inserted periodically in the receiving frames, which does not apply to code-aided frame synchronization. On the other hand, the computation of our algorithm is not compared with other algorithms, which will be further studied.

Author Contributions: Conceptualization, H.Y.; methodology, J.C.; software, H.Y.; supervision, Z.H.; validation, S.L.; writing (original draft), H.Y.; writing, review and editing, S.L. All authors have read and agreed to the published version of the manuscript.

Funding: This research was supported by the National Natural Science Foundation of China (General Program), Grant No. 51575517.

Acknowledgments: The authors would like to acknowledge Qiquan Zheng for providing high-performance computer as a simulation platform, and the authors also would like to thank editors and anonymous reviewers for giving valuable suggestions that have helped to improve the quality of the manuscript.

Conflicts of Interest: The authors declare no conflict of interest.

References

1. Wozencraft, J.M.; Jacobs, I.M. *Principles of Communication Engineering*; Wiley: New York, NY, USA, 1965.
2. *TM Synchronization and Channel Coding*; Blue Book; Issue 1; CCSDS: Washington, DC, USA, 2003.
3. Mehdi, T.; Hamidreza, M. Reverse Engineering of Communications Networks: Evolution and Challenges. *arXiv*, 2017; arXiv:1704.05432.
4. Guangzu, L.; Jianxin, W.; Tian, B. Frame Detection Based on Cyclic Autocorrelation and Constant False Alarm Rate in Burst Communication Systems. *China Commun.* **2015**, *12*, 55–63.
5. Barker, R.H. Group Synchronization of Binary Digital Systems. *Commun. Theory* **1953**, *20*, 273–287.
6. Choi, Z.Y.; Lee, Y.H. On the Use of Double Correlation for Frame Synchronization in the Presence of Frequency Offset. In Proceedings of the 1999 IEEE International Conference on Communications, Vancouver, BC, Canada, 6–10 June 2019; pp. 958–962.
7. Choi, Z.Y.; Lee, Y.H. Frame synchronization in the presence of frequency offset. *IEEE Trans. Commun.* **2002**, *50*, 1062–1065. [CrossRef]
8. Massey, J.L. Optimum Frame Synchronization. *IEEE Trans. Commun.* **1972**, *20*, 115–119. [CrossRef]
9. Nielsen, P.T. Some Optimum and Suboptimum Frame Synchronizers for Binary Data in Gaussian Noise. *IEEE Trans. Commun.* **1973**, *21*, 770–772. [CrossRef]
10. Lui, G.L.; Tan, H.H. Frame synchronization for Gaussian channels. *IEEE Trans. Commun.* **1987**, *35*, 818–829.
11. Robertson, P. Maximum likelihood frame synchronization for flat fading channels. In Proceedings of the 1992 IEEE International Conference on Communications, Chicago, IL, USA, 14–18 June 1992; pp. 1426–1430.
12. Wang, Y.; Shi, K.; Serpedin, E. Continuous Mode Frame Synchronization for Frequency Selective Channels. *IEEE Trans. Commun.* **2004**, *53*, 865–871. [CrossRef]
13. Kopansky, A.; Bystrom, M. Frame Synchronization for Noncoherent Demodulation on Flat Fading Channels. In Proceedings of the 2000 IEEE International Conference on Communications, New Orleans, LA, USA, 18–22 June 2000; pp. 312–316.
14. Chiani, M.; Martini, M. Optimum synchronization of frames with unknown, variable lengths on Gaussian channels. In Proceedings of the IEEE Global Telecommunications Conference, Dallas, TX, USA, 29 November–3 December 2004; Volume 6, pp. 4087–4091.
15. Chiani, M.; Martini, M. Practical frame synchronization for data with unknown distribution on AWGN channels. *IEEE Commun. Lett.* **2005**, *9*, 456–458. [CrossRef]
16. Chiani, M.; Martini, M. On sequential frame synchronization in AWGN channels. *IEEE Trans. Commun.* **2006**, *54*, 339–348. [CrossRef]
17. Kim, P.; Pedone, R.; Villanti, M.; Vanelli-Coralli, A.; Corazza, G.E.; Chang, D.I.; Oh, D.G. Robust frame synchronization for the DVB-S2 system with large frequency offsets. *Int. J. Satell. Commun. Netw.* **2009**, *27*, 35–52. [CrossRef]
18. Pedone, R.; Villanti, M.; Vanelli-Coralli, A.; Corazza, G.; Mathiopoulos, P.T. Frame Synchronization in Frequency Uncertainty. *IEEE Trans. Commun.* **2010**, *58*, 1235–1246. [CrossRef]
19. Bai, Y.; Yang, J.X.; Zhang, Y. A Recognition Method of m-sequence Synchronization Codes using Higher-order statistical Processing. *J. Electrinics Inf. Technol.* **2012**, *34*, 33–37. (In Chinese) [CrossRef]
20. Bai, Y.; Yang, J.X.; Zhang, Y. Recognition Method of Fame Synchronization Codes based on Relativity Filter and Hadamard transformation Algorithm. *J. Detect. Contol.* **2011**, *33*, 69–72. (In Chinese)
21. Naseri, A.; Salemm, J.; Hajimohammadii, M. Improving the blind recovering algorithm and synchronization word in satellite communication systems. *Indian J. Sci. Res.* **2014**, *1*, 15–19.
22. Qin, J.Y.; Huang, Z.P. Novel Blind Recognition Algorithm of Frame Synchronization Words Based on Soft-Decision in Digital Communication Systems. *PLoS ONE* **2015**, *10*, e0132114. [CrossRef]
23. Guo, F.K.; Wang, M. Research on Detection of Frame Head on Fixed Frame Length. *J. Signal Inf. Process.* **2014**, *44*, 33–36. (In Chinese)
24. Xu, Y.Y.; Zhong, Y.; Huang, P.Z. An Improved Blind Recognition Algorithm of Frame Parameters Based on Self-Correlation. *Information* **2019**, *10*, 64. [CrossRef]
25. Imad, R.; Sicot, G.; Houcke, S. Blind frame synchronization for error correcting codes having a sparse parity check matrix. *IEEE Trans. Commun.* **2009**, *57*, 1574–1577. [CrossRef]

26. Xia, T.; Wu, H. Joint Blind Frame Synchronization and Encoder Identification for Low-Density Parity-Check Codes. *IEEE Commun. Lett.* **2014**, *18*, 353–355. [CrossRef]

27. Zhou, J.; Huang, Z.; Su, S. Blind Frame Synchronization of Reed-Solomon Coded Optical Transmission Systems. *OPTIK* **2013**, *124*, 998–1002.

28. Qi, Y.; Wang, B.; Rong, M.; Liu, T. Comments on "Theoretical Analysis of a MAP Based Blind Frame Synchronizer. *IEEE Trans. Wirel. Commun.* **2011**, *10*, 3127–3132.

29. Hough, P.V.C. A Method and Means for Recognizing Complex Patterns. U.S. Patent 3,069,654, 18 December 1962.

30. Mukhopadhyay, P.; Chaudhuri, B. A survey of Hough Transform. *Pattern Recognit.* **2014**, *48*, 993–1010. [CrossRef]

31. Duda, R.O.; Hart, P.E. Use of the Hough transformation to detect lines and curves in pictures. *Commun. ACM* **1972**, *15*, 11–15. [CrossRef]

32. Imad, R.; Houcke, S. Theoretical Analysis of a MAP Based Blind Frame Synchronizer. *IEEE Trans. Wirel. Commun.* **2009**, *8*, 5472–5476. [CrossRef]

Article

Peak-to-Average Power Ratio Reduction of Carrier-Suppressed Optical SSB Modulation: Performance Comparison of Three Methods

K. I. Amila Sampath [1,*] , **Katsumi Takano** [2] **and Joji Maeda** [1]

1   Department of Electrical Engineering, Faculty of Science and Technology, Tokyo University of Science, Noda 278-8510, Japan; joji.maeda@rs.tus.ac.jp
2   Department of Informatics and Electronics, Faculty of Engineering, Yamagata University, Yonezawa 992-8510, Japan; ktakano@yz.yamagata-u.ac.jp
*   Correspondence: amila.sampath@rs.tus.ac.jp

Abstract: We compare the performances of three previously proposed methods to reduce the peak-to-average power ratio (PAPR) of the carrier-suppressed optical single-sideband (OSSB-SC) signal. PAPR of OSSB-SC signal becomes high due to the peaky Hilbert-transformed signal which is used for spectral suppression. Nonlinear phase shifts induced by high PAPR degrade OSSB-SC signal during fiber transmission. Previously, we proposed peak folding, peak clipping, and high-pass Hilbert transform methods to reduce the PAPR of OSSB-SC modulation. In this study, we numerically compare the effectiveness of proposed methods in a 10 Gbit/s non-return-to-zero (NRZ)-coded 100-km single-channel transmission link. Due to the reduced PAPR, peak folding and peak clipping can increase the self-phase modulation (SPM) threshold of the studied system by 2.40 dB and 2.63 dB respectively. The high-pass Hilbert transform method improves the SPM threshold by more than 9 dB.

Keywords: optical communications; BPSK-VSB; BPSK-SSB; fiber transmission

Citation: Sampath, K.I.A.; Takano, K.; Maeda, J. Peak-to-Average Power Ratio Reduction of Carrier-Suppressed Optical SSB Modulation: Performance Comparison of Three Methods. *Photonics* **2021**, *8*, 67. https://doi.org/10.3390/photonics8030067

Received: 20 January 2021
Accepted: 23 February 2021
Published: 26 February 2021

Publisher's Note: MDPI stays neutral with regard to jurisdictional claims in published maps and institutional affiliations.

1. Introduction

Tremendous amounts of data traffic being added to short-reach networks are expediting data rate scaling of 100 to 400 G and beyond [1–3]. Intensity modulation-direct detection (IM-DD) transmission is preferred for short-reach links because of its simplicity and cost-effectiveness [4–6]. However, to meet the capacity requirements, it is becoming obvious that more degree of freedom is required in short-reach links. Driven by emerging capacity hungry applications, digital coherent transmission with the high-order degree of freedom using polarization and wavelength-multiplexing is gaining attention in short-reach links, regardless of the cost [7–9].

Recently, optical single-sideband (OSSB) modulation with direct-detection took the attention of researchers as a cost-effective solution for increasing the capacity of short-reach links [10–12]. In C-band transmission, OSSB modulation can tolerate signal distortions induced by chromatic dispersion because of the single-sided spectrum [13,14]. Moreover, the phase information of the transmitted optical signal is preserved even after direct detection. Therefore, electrical dispersion compensation can be done in the receiver [15,16]. However, signal-signal beat interference (SSBI) caused by square-law detection of the OSSB signal degrades the received signal. Several methods have been studied to reduce or eliminate the SSBI of the direct-detected OSSB signal [17–20]. Kramers-Kroning (KK) relation-based receiver outperforms all the SSBI-cancellation schemes proposed so far [20]. Nevertheless, the KK receiver requires faster digital signal processing (DSP) and a high carrier-to-signal power ratio (CSPR) [21,22].

Even though KK reception is applicable for both double sideband (optical double-sideband signal+ optical tone at the edge of the spectrum) and SSB (OSSB + optical tone)

transmission [23], SSB transmission is preferred to save the DSP bandwidth at the receiver [24,25]. Equivalently, the addition of optical tone at the receiver is preferred because it allows polarization division multiplexing (PDM) while increasing the transmission power efficiency [26]. In this manner, adding an optical tone to transmitted carrier-suppressed OSSB signal (OSSB-SC) at the receiver can overcome two major technical challenges of the KK receiver.

Despite OSSB-SC transmission's capability of overcoming the above-mentioned challenges, inherited high peak-to-average power ratio (PAPR) of OSSB-SC signal introduces two new challenges; fiber nonlinearity-based signal distortions and requirement of high tone power at the receiver to meet the minimum phase condition. OSSB-SC signal consists of high peaks in the optical waveform. This is because of the Hilbert-transform pair relation of in-phase and quadrature-phase components of the OSSB signal [27].

Paying attention to the fiber nonlinearity-based signal distortions, the authors previously proposed three methods for alleviating PAPR of OSSB-SC signal, namely, peak folding using optical modulator nonlinearity [28], peak clipping [29], and high-pass Hilbert transforming [30]. The proposed methods use transmitter side digital processing or nonlinear modulation characteristics of the LN (Lithium niobate: $LiNbO_3$) IQ modulator. Extending our previous work, after brief introductions of operation principles, we compare the effectiveness of the above three methods in a 100-km transmission system in this paper. To clarify the transmission performance improvements by PAPR reduction, we restrict ourselves to single-channel single-polarization binary phase-shift keying (BPSK) transmission. To avoid the effect of the receiver DSP parameters on the proposed methods, we use ideal coherent detection to recover the transmitted signal. Because PAPR is closely related to self-phase modulation (SPM)-based signal distortions, the SPM threshold is used as a figure of merit. PAPR reductions by proposed methods are compared and changes of modulated signal spectra during the PAPR reduction process are discussed.

The rest of this paper is organized as follows; in Section 2, the principal of the phase-shift method OSSB-SC signal generation and characteristics of OSSB-SC signal are discussed. Section 3 describes the PAPR reduction of the OSSB-SC signal by folding the peaks of the Hilbert-transformed signal. PAPR reduction by peak clipping is discussed in Section 4. High-pass Hilbert transform-based PAPR reduction is discussed in Section 5. The effectiveness of three PAPR reduction techniques in fiber transmission is compared in Section 6. In Section 7, we discussed the reported results in detail before conclusions are drawn.

2. Phase-Shift Method OSSB-SC Signal and Its PAPR

Figure 1a shows the schematic of the optical IQ modulator-based optical SSB transmitter. IQ modulator is composed of two sub-Mach-Zehnder interferometers (sub-MZIs) in a dual parallel structure. To generate OSSB-SC signal, baseband signal $V_B(t)$ and its Hilbert transform $V_H(t)$ are used to drive the two sub-MZIs which are biased at their transmission null points. Hilbert transform is defined as in Equation (1).

$$H(\omega) = \begin{cases} -j\mathrm{sgn}(\omega) & (\omega \neq 0) \\ 0 & (\omega = 0) \end{cases} \tag{1}$$

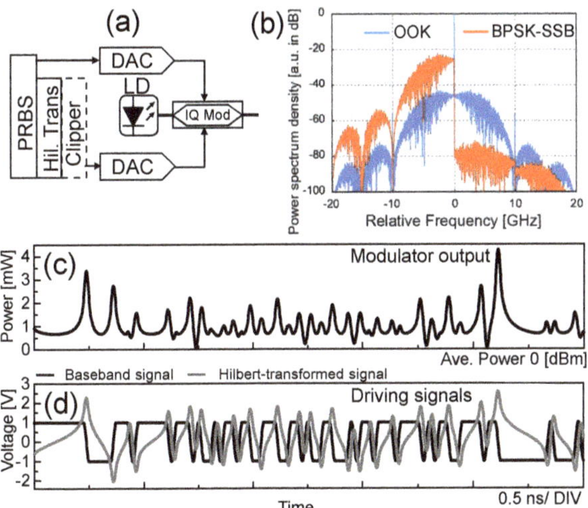

Figure 1. (**a**) IQ modulator based OSSB-SC transmitter; (**b**) spectra of optical BPSK-SSB signal and OOK modulated signal; (**c**) temporal waveform of optical BPSK-SSB modulator output power; (**d**) temporal waveforms of modulator driving signals, PRBS: pseudo-random binary sequence generator, DAC: digital-to-analog converter, OOK: on-off keying, BPSK-SSB: binary phase-shift keying-single sideband.

Here, sgn($\cdot$) denotes the signum function. ω is the angular frequency. Hilbert transform creates a π-phase difference between the upper and lower frequency components of the baseband spectrum separated by the center frequency. By orthogonally combining the output light of two sub-MZIs, a side-band suppressed signal is generated. Sideband suppression is achieved due to the π-phase difference between the spectral sidebands. Therefore, this modulation is named the phase-shift method [31].

We define modulation depth as the ratio of the peak voltage of the baseband signal ($V_{\text{B-Peak}}$) to the half-wave voltage (V_π) of the IQ modulator. Figure 1b compares the optical power spectra of OSSB-SC modulated and optical on-off keying (OOK) modulated 10 Gbit/s non-return-to-zero (NRZ)-coded sequence when modulation depth is 0.1. Over 40-dB spectral sideband suppression can be observed in the OSSB-SC spectrum compared to that of the intensity-modulation. The output optical waveform and the two driving signal waveforms of the IQ modulator are depicted in Figure 1c,d, respectively. Here, the baseband signal is an NRZ-coded binary sequence. Peaks appear in the Hilbert-transformed waveform following the transmission points of the baseband signal between marks and spaces. The height of the peaks in the Hilbert-transformed waveform depends on the transfer function of the Hilbert transform [27] and the bit pattern of the baseband signal. As can be noticed by comparing the driving signals and the modulator output optical waveforms, peaks of the Hilbert-transformed component cause peaks in the modulator output waveform. Subsequently, the PAPR of the modulator output increases. High PAPR of the optical output leads to signal distortions at the receiver due to nonlinear phase-shifts caused by SPM during the transmission.

3. Peak Folding Using LN Modulator Non-Linearity

3.1. Principle

The operation condition of the IQ modulator for OSSB-SC signal generation is illustrated in Figure 2. To modulate the amplitude of the optical carrier, two sub MZIs of the IQ modulator are biased into their transmission null points. In conventional OSSB-SC modulation schemes, electrical input signals are driven within the range of $2V_\pi$. However, peaks of Hilbert-transformed signal which causes high PAPR can be suppressed using the sinusoidal shape of the MZI modulation curve [28]. Portions of the Hilbert transformed signal which surpass the range of $2V_\pi$ are folded back. One can use this peak folding property of the IQ modulator to suppress the peaks of the Hilbert-transformed signal and to reduce the PAPR of the OSSB-SC signal.

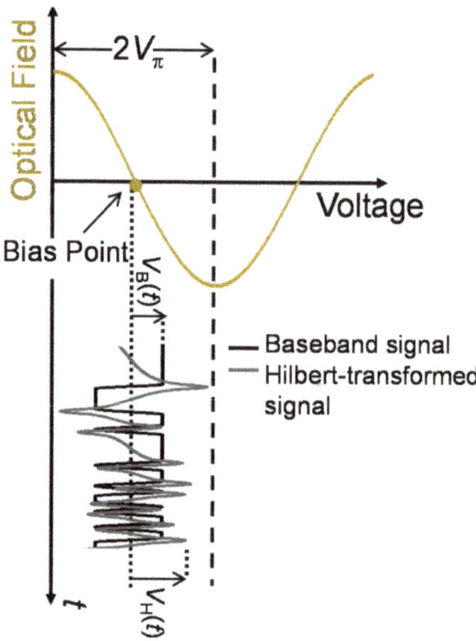

Figure 2. The operation condition of IQ modulator.

Because the peak-to-peak voltage of the Hilbert transformed signal ($V_{\text{pp-H}}$) is about 2.5 times that of the baseband signal, $V_{\text{pp-H}}$ exceeds $2V_\pi$ range for modulation depths larger than 0.4. Peak folding was implemented by increasing the voltage of the Hilbert-transformed signal component. Accordingly, the modulation depth was increased.

3.2. Characteristics of Peak-Folded OSSB-SC Signal

We numerically investigated the OSSB-SC signal generated using a peak-folded Hilbert-transformed component. The OSSB-SC transmitter model is depicted in Figure 1a. A continuous-wave (CW) light of 1552.5 nm from a laser diode (LD) was sent to the IQ modulator. Optical BPSK-SSB signal was generated by driving two MZIs of IQ modulator by 10 Gbit/s NRZ coded (pseudo-noise (PN)-stage 10) baseband signal generated at a random bit sequence generator and its Hilbert transform. Peak folding was achieved for the values of $V_H(t)$ greater than V_π. Modulator output power also increases with modulation depth. However, for a better comparison of waveform changes, the average modulator output power was kept constant by adjusting the LD power when the modulation depth was changed. We assumed identical half-wave voltages for the two-sub MZIs.

Figure 3a presents modulator output waveforms when the modulation depth was 0.1, 0.6, and 1. When the modulation depth is 0.1, sharp peaks can be noticed in the modulator output waveform. This is because of the peaks of the Hilbert-transformed waveform which have transferred linearly from the electrical domain to the optical domain at the linear region of the modulation curve when the modulation depth was small. The peaks of the modulator output start to shrink when the modulation depth is increased. For the modulation depths greater than 0.4, $V_H(t)$ becomes larger than V_π. Hence the peaks of the Hilbert transformed waveform are folded back by the modulation curve. This peak folding back of the Hilbert-transformed waveform appears as peak shrinking in the modulator output waveform. The minimum PAPR of 1.52 was achieved by peak folding at a modulation depth of 1.0.

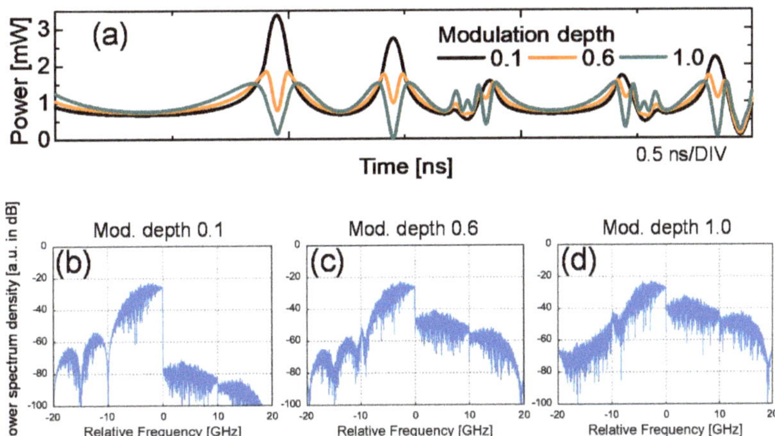

Figure 3. Waveforms and power spectra of peak-folded OSSB-SC signal, (**a**) modulator output waveforms at modulation depths of 0.1, 0.6, and 1.0, (**b–d**) power spectra of the modulator output signal at modulation depths of 0.1, 0.6, and 1.0 respectively.

Spectra of the modulator output signal are depicted for comparison in Figure 3b–d when the modulation depth was 0.1, 0.6, and 1.0, respectively. The power spectrum density of the suppressed sideband has increased with the modulation depth. In this study, we define the sideband suppression ratio (SSR) as the difference of peak power spectral densities of the suppressed sideband and the unsuppressed sideband as shown in the inset of Figure 4. SSR of 48.7 dB at the modulation depth of 0.1 increased to 17.6 dB when the modulation depth was changed to 0.6.

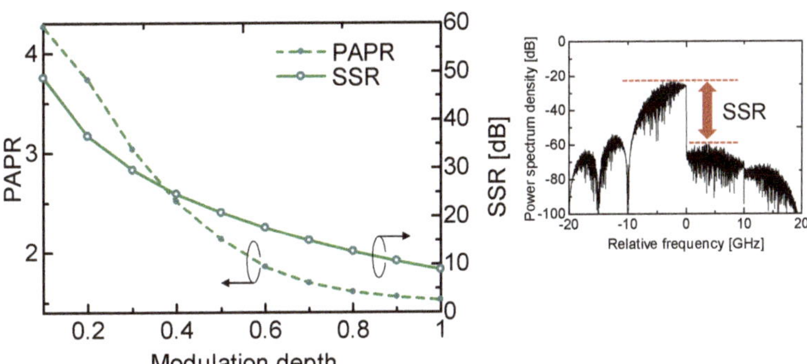

Figure 4. PAPR and SSR characteristics of peak-folded OSSB-SC signal, inset; definition of sideband suppression ratio (SSR).

PAPR and SSR variations of the modulator output signal with the modulation depth are depicted in Figure 4. By increasing the modulation depth from 0.1 to the maximum modulation depth of 1, PAPR is reduced from 4.26 to 1.52. During the increase of modulation depth, the Hilbert-transformed waveform was degraded by peak folding. Peak-folded Hilbert-transformed component leads to harmonics in the optical spectrum because of the nonlinear modulation characteristics of the IQ modulator. Subsequently, an increase in suppressed sideband power is observed causing degradation of SSR. However, it is noticeable that SSR greater than 20 dB is achievable for the modulation depth of 0.5 where PAPR becomes 2.13.

4. Peak Clipping

4.1. Principle

Another approach for PAPR reduction of the OSSB-SC signal is clipping the peaks of the Hilbert-transformed signal in the electrical domain [29]. Peak clipping was proposed and extensively studied as a PAPR reduction method for wireless orthogonal frequency-division multiplexing (OFDM) transmission [32,33]. Peak clipping of OFDM results in direct signal-amplitude distortions since peaks of the OFDM signal are superpositions of OFDM subcarriers where the data is encoded [32]. However, since the peaks of modulator output optical signal are resulted by peaky Hilbert-transformed component in OSSB-SC modulation, the peak clipping of Hilbert-transformed signal does not cause direct harm to baseband signal amplitude.

Peak clipping of Hilbert transformed signal can be implemented using a clipper circuit which was introduced to the transmitter after Hilbert transformer (Figure 1a). Peak clipped Hilbert-transformed waveform is illustrated in Figure 5 with a pink line. The maximum value of clipping voltage V_{CL} is restricted to V_π by the sinusoidal shape modulation curve of MZI. Here $V_{\text{pp-B}}$ denotes peak-to-peak voltages of the baseband signal. $V_{\text{B-Peak}}$ is the peak voltage of the baseband signal.

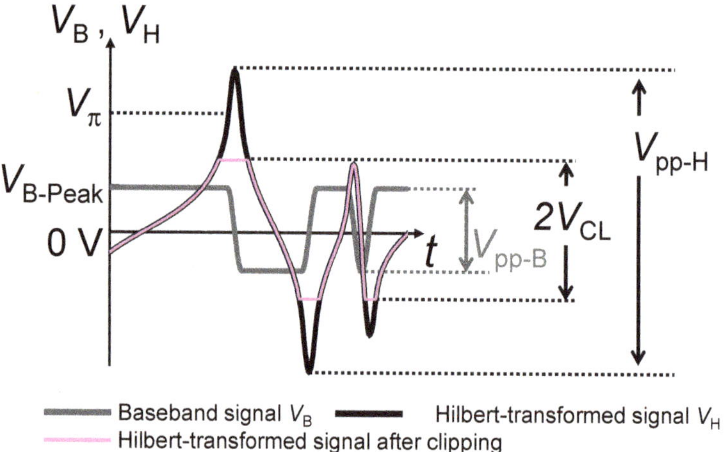

Figure 5. Peak clipping of Hilbert-transformed signal.

4.2. Characteristics of Peak-Clipped OSSB-SC Signal

Properties of the peak clipped OSSB-SC signal were analyzed using the transmitter model in Figure 1a. The effect of peak clipping was studied for a fixed clipping voltage [29] and variable clipping voltages [34]. The same driving signals described in 3.2 were used to drive the modulator. The Hilbert-transformed signal component which produces peaks in the modulator output waveform was clipped using the clipper circuit before connecting to the modulator.

4.2.1. Fixed Clipping Voltage

When the clipping voltage is fixed, the clipping amount of Hilbert-transformed waveform depends on the values of modulation depth and the clipping voltage V_{CL}. Here we discuss the peak clipped OSSB-SC signal when V_{CL} equals V_π. The clipping amount was adjusted by changing the modulation depth by varying $V_{\text{B-Peak}}$. An increase of $V_{\text{B-Peak}}$ increases $V_{\text{pp-H}}$ accordingly. When the modulation depth is greater than 0.4, $V_{\text{pp-H}}$ exceeds $\pm V_\pi$ range. Because V_{CL} equals V_π, Hilbert transformed waveform is clipped when the amplitude surpasses the range of $\pm V_\pi$. For modulation depths greater than 0.4, Hilbert transformed waveform exceeds $\pm V_\pi$ range. Consequently, the upper MZI of the IQ modulator was driven by the peak-clipped Hilbert-transformed waveform.

In both fixed and variable clipping voltage studies, the modulator output power was unchanged. We adjusted the LD output power in the case of modulation depth or clipping voltage was changed.

Modulator output waveforms of peak-clipped OSSB-SC signal are compared in Figure 6a. The waveforms of Figure 6a are calculated for modulation depths of 0.1, 0.4, 0.6, and 1.0 when the average modulator output power was 0 dBm. Because V_{CL} was set to V_π, there were no peak clippings when the modulation depth was 0.1 where $V_{\mathrm{pp\text{-}H}}$ nearly equals $0.5 V_\pi$. As a consequence, high peaks appear in modulator output intensity waveform as mentioned in Section 3.2. Peak reduction can be seen at the modulation depth of 0.4 where the peak clipping has just started. The cause of peak reduction up to modulation depth of 0.4 was the nonlinearity of the modulator transfer function. When the amplitude of the modulator input signal becomes sufficiently large with increasing modulation depth, the effect from the sinusoidal shape of the transfer function of MZI appears as peak folding of modulated signals. Since the Hilbert-transformed signal has a larger amplitude it is affected by the nonlinearity of the modulator first and reduces the peaks of the modulator output signal [28]. For modulation depths greater than 0.4, $V_{\mathrm{pp\text{-}H}}$ exceeded the $\pm V_\pi$ range and peak clipping was implemented. Consequently, peak suppression of modulated output waveform can be seen which results in a reduction of PAPR.

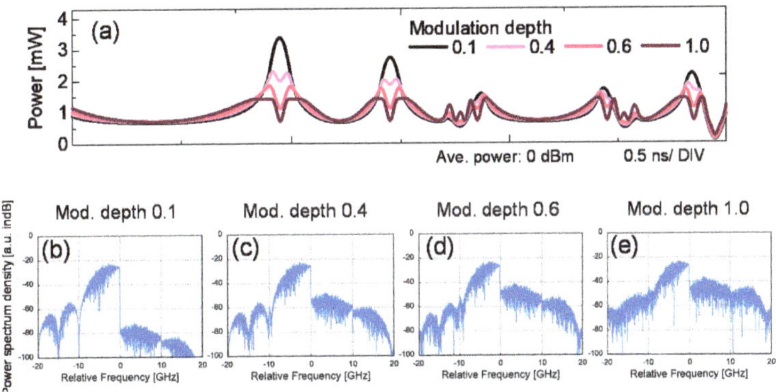

Figure 6. Peak clipped OSSB-SC signal ($V_{\mathrm{CL}} = V_\pi$), (**a**) modulator output intensity waveform at modulation depths of 0.1, 0.4, 0.6 and 1.0, (**b–e**) power spectra of modulator output signals at modulation depths of 0.1, 0.4, 0.6 and 1.0 respectively.

Figure 6b–e show the power spectra of the modulator output signal when the modulation depths were 0.1, 0.4, 0.6, and 1.0, respectively. Compared with Figure 6b where peak clipping was not implemented, a rise in suppressed sideband power can be noticed in peak-clipped signal spectra. Both peak clipping and the nonlinearity of the IQ modulator contribute to this spectral degradation [29].

PAPR and SSR variations of the peak-clipped OSSB-SC signal are depicted in Figure 7. With increasing modulation depth, the clipping amount of the Hilbert-transformed signal increases consequently reducing the PAPR. Due to peak clipping, the value of SSR decreases. SSR becomes 13.57 dB at the modulation depth of 1.0 where the minimum PAPR of 1.41 was achieved. PAPR and SSR characteristics of peak-folded OSSB-SC signal are also shown for comparison. PAPR, SSR values of both peak-folded and peak-clipped signals resemble lower modulation depths. Compared with peak folding, peak clipping reduces waveform degradations of Hilbert-transformed signal at high modulation depths and thereby reducing harmonics generated during the modulation. Subsequently, peak- clipped

OSSB-SC signal shows lower PAPR and higher sideband suppression relative to the peak-folded signal. Peak clipping improves PAPR by 6.9% and SSR by 4.7 dB at the modulation depth of 1.0.

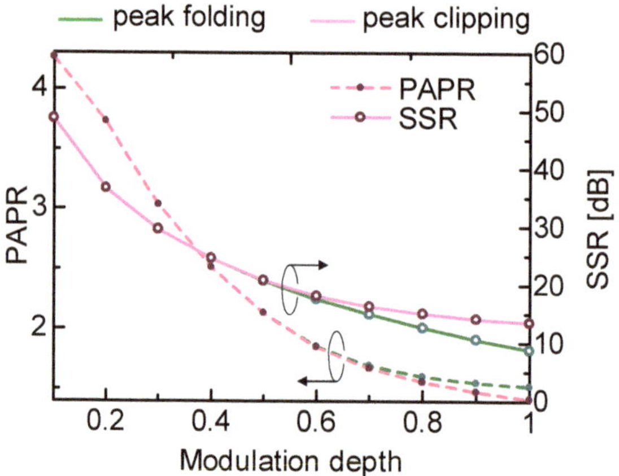

Figure 7. PAPR and SSR of peak clipped OSSB-SC signal.

4.2.2. Variable Clipping Voltage

Then the effect of the clipping voltage was studied while varying the clipping voltage when the modulation depth was fixed. We compared the characteristics of peak-clipped OSSB-SC signal when the modulation depth was 0.1 and 0.4. Modulator output waveforms and their spectra at modulation depths of 0.1 and 0.4 are depicted in Figure 8 for varied clipping voltages. To check the effect of variable clipping voltage, clipping ratio r is defined as

$$r = \frac{V_{CL}}{V_{B-Peak}} \tag{2}$$

The voltages of modulator input electrical signals vary with the modulation depth. However, the waveforms and spectra of Figure 8 are compared when the amount of clipping of the Hilbert-transformed waveform relative to the baseband signal is equal. The average modulator output power was 0 dBm in each case.

For larger r, modulator output waveforms show peaks in both 0.1 and 0.4 modulation depths. Because the clipping amount increases with decreasing r, peaks start to shrink. Waveforms of both modulation depths start to resemble when r becomes smaller. Suppression of peaks in modulator output waveform can be observed for the modulation depth of 0.4 even for smaller values of r. This difference of peak powers of two modulation depths at higher values of r is due to the nonlinearity of the IQ modulator.

The effect of the modulator nonlinearity can be also seen in the power spectra of peak clipped OSSB-SC signal. For greater modulation depths, the power spectrum density of the suppressed sideband becomes high. Harmonics generated due to the nonlinearity of the modulation curve of the IQ modulator increases the power of the suppressed sideband. Spectra and modulator output intensity waveforms of the modulation depth of 0.1 mimic those of the modulation depth of 0.4 since the baseband signal component of the OSSB-SC signal becomes dominant with decreasing r.

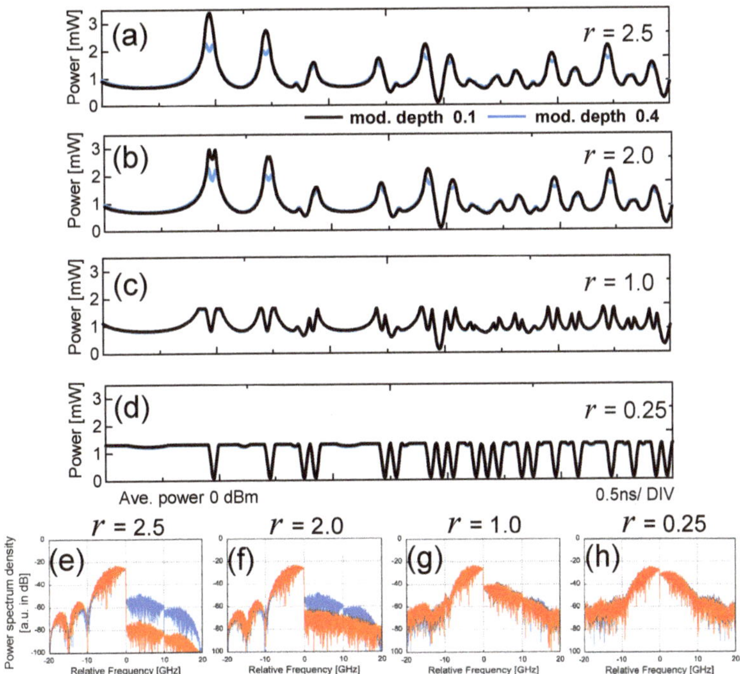

Figure 8. Peak-clipped OSSB-SC signal with variable clipping voltage, (**a–d**) modulator output intensity waveforms at *r* = 2.5, *r* = 2.0, *r* = 1.0, *r* = 0.25 respectively. (**e–h**) power spectra of the waveforms of (**a–d**) respectively.

PAPR and SSR variations of the peak-clipped OSSB-SC signal are illustrated in Figure 9. PAPR of 4.14 was noticed for the modulation depth of 0.1 when *r* is 2.5. As a result of peak folding, PAPR was reduced to 2.51 at the modulation depth of 0.4. Because of suppressed peaks by peak clipping, PAPR reduces gradually with *r*. Minimum PAPR of 1.30 was achieved for *r* < 0.5. SSR becomes small with decreasing *r* for both modulation depths. For large *r*, the SSR becomes smaller for the modulator output signal at a modulation depth of 0.4, because of harmonics generated due to the nonlinearity of the modulator. Because peak clipping of the Hilbert-transformed component also contributes to spectral degradation by adding harmonics, the SSR becomes small for smaller *r*. It is noteworthy that the SSR for both modulation depths coincides when *r* is decreased.

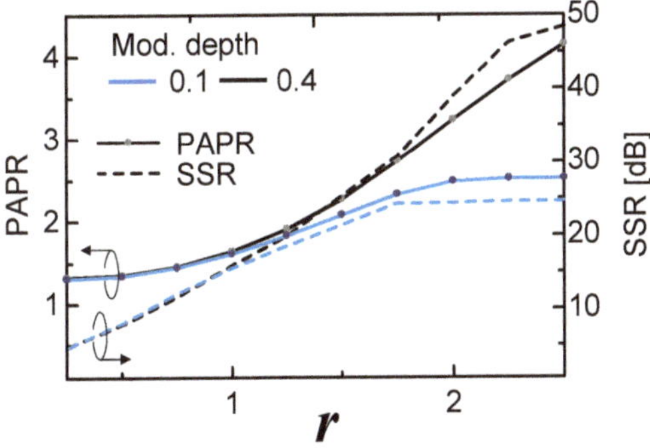

Figure 9. PAPR and SSR of peak clipped OSSB-SC signal for variable clipping voltage.

5. High-Pass Hilbert Transform

PAPR of OSSB-SC signal also can be reduced by manipulating the spectrum of peaky Hilbert-transformed components [30]. Because the shape of the waveform is largely determined by low-frequency components, peaks of the Hilbert-transformed waveform can be suppressed by reducing their power. To suppress the low-frequency components power of Hilbert-transformed signal, we modify the flat amplitude response of Hilbert transform to a high-pass response. The modified transfer function of the Hilbert transform is as follows

$$H_H(\omega) = \begin{cases} -j\,\mathrm{sgn}(\omega) & (|\omega| > 2\omega_c) \\ -j\,\mathrm{sgn}(\omega)\sin\{\pi(|\omega|/4\omega_c)\} & (|\omega| \le 2\omega_c) \end{cases} \tag{3}$$

Here, ω_c is the cut-off frequency. Amplitude and phase responses of modified Hilbert transform are compared with those of ideal Hilbert transform in Figure 10a–c. Because of the high-pass amplitude response, we name H_H as high-pass Hilbert transform. The high-pass Hilbert-transformed waveform at a cut-off frequency of 3 GHz is portrayed in Figure 10d and compared with 10 Gbit/s NRZ-coded baseband signal and its ideal Hilbert-transformed waveforms. The high-pass Hilbert transform reduces the peaks of the Hilbert-transformed waveform. The peak-to-peak voltage of the high-pass Hilbert-transformed waveform reached that of the baseband signal at the cut-off frequency of 3 GHz.

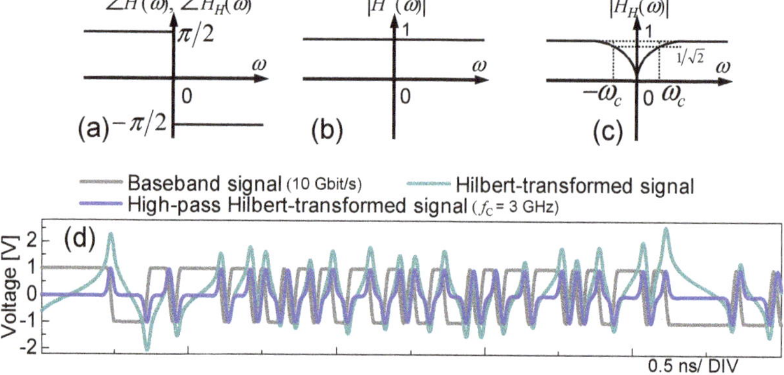

Figure 10. (**a**) phase response of $H(\omega)$ and $H_H(\omega)$, (**b**) amplitude response of Hilbert transform $H(\omega)$, (**c**) amplitude response of high-pass Hilbert transform $H_H(\omega)$, (**d**) 10 Gbit/s NRZ-coded baseband signal and its Hilbert-transformed waveforms ($f_c = 3$ GHz of $H_H(\omega)$).

Figure 11 shows the relation of driving signals voltage ratio and the upper limit of the modulation depth of high-pass Hilbert transformer-based OSSB-SC signal. The baseband signal is a 10 Gbit/s NRZ-coded sequence. With increasing cut-off frequency, peaks of Hilbert-transformed signal component reduce. This reduction of Hilbert-transformed signal amplitude can be observed as a reduction of the driving signal peak-to-peak voltage (V_{pp}) ratio which is defined as the ratio of $V_{pp\text{-}H}$ to $V_{pp\text{-}B}$. The V_{pp} ratio reaches unity at the cut-off frequency of 3 GHz. Further increase of the cut-off frequency results in V_{pp} ratios smaller than 1 because the peak-to-peak voltage of the Hilbert-transformed signal becomes smaller than that of the baseband signal. Because the V_{pp} ratio becomes unity for cut-off frequencies greater than 3 GHz, maximum modulation depth can be achieved for cut-off frequencies greater than 3 GHz. Different from peak folding and peak clipping, the high-pass Hilbert transform allows modulation at higher modulation depths with lower spectral degradations due to the decrease of $V_{pp\text{-}H}$.

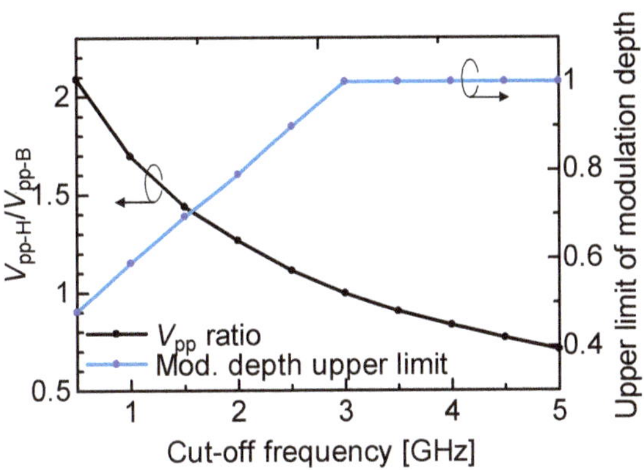

Figure 11. Driving signals peak-to-peak voltage (V_{pp}) ratio and the modulation depth's upper limit of the high-pass Hilbert transformer based OSSB-SC signal.

PAPR and optical bandwidth of the modulator output signal are plotted against the cut-off frequency and presented in Figure 12. Peak reduction of Hilbert-transformed signal results reduced PAPR of the modulator output signal. PAPR reaches unity around the cut-off frequency of 3 GHz.

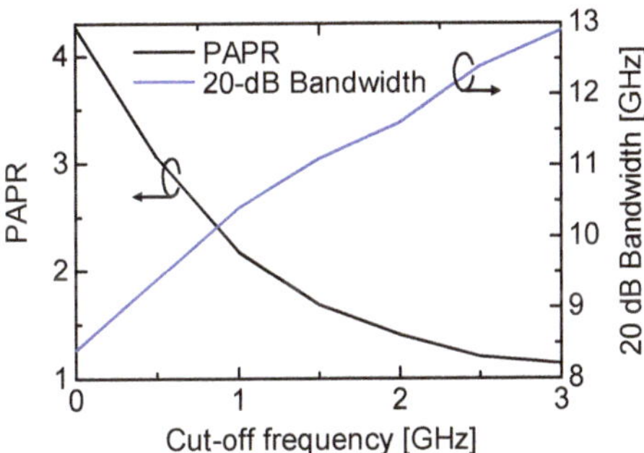

Figure 12. PAPR and 20-dB optical bandwidth of the modulator output signal.

Because of the filtering of Hilbert-transformed signal, residual bandwidth of modulator output signal increases with increasing cut-off frequency. Spectral changes of high-pass Hilbert transform-based OSSB-SC signal are evaluated using 20-dB bandwidth of the optical spectrum. 20-dB bandwidth is defined as the spectral bandwidth at where the power spectral density becomes -20 dB relative to the maximum power spectral density of the modulator output signal spectrum. Because high-pass Hilbert transform filters out the lower frequencies of the Hilbert-transformed signal spectrum, sideband suppression deteriorates. Consequently, the 20-dB bandwidth increases almost linearly with the cut-off frequency. However, even at a cut-off frequency of 3 GHz, 20-dB bandwidth remains within 65% of double sideband modulation bandwidth giving spectral efficiency of 0.77 b/s/Hz.

6. Fiber Transmission of PAPR Reduced OSSB-SC Signal

To study the effect of PAPR reduction on transmission characteristics of the OSSB-SC signal by the proposed methods, a 100-km transmission simulation was carried out. OSSB-SC signal was generated as described in Sections 3–5. The modulator output signal was launched into the fiber link shows in Figure 13 and transmitted.

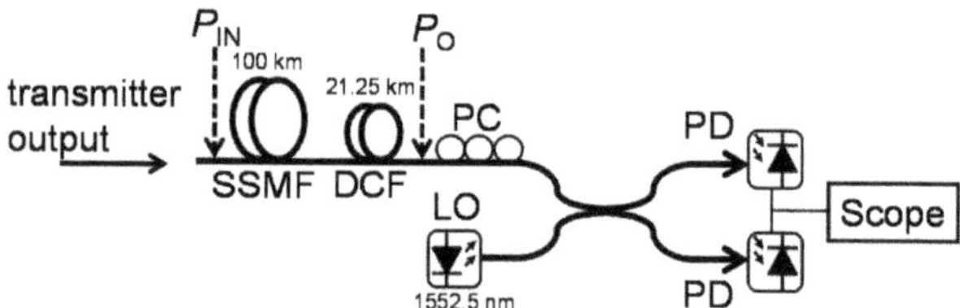

Figure 13. Simulated fiber transmission setup.

We assumed ideal phase-diversity homodyne detection at the receiver instead of KK relation-based direct detection. This allows us to clarify the effect of PAPR reduction on nonlinear signal distortions during transmission, without bothering to optimize the DSP parameters required for the KK receiver, which is out of the scope of this paper. For the same reason, we used dispersion compensating fiber (DCF) in our simulation model to compensate fiber dispersion instead of frequency-domain equalization (FDE). Chromatic dispersion of standard single-mode fiber (SSMF) was compensated using 21.5-km long DCF. We found no waveform degradations during the transmission of DCF. Because of the superior sensitivity characteristics of coherent detection [23,35,36], we did not employ any optical amplifiers in our transmission simulations. Here P_{IN} and P_O represent average fiber input and output power, respectively. We simulated fiber transmission by solving the nonlinear Schrödinger equation using the split-step Fourier method [37]. Fiber parameters used in our calculations are given in Table 1. At the receiver, the transmitted signal was detected using a phase-diversity homodyne detector. The receiver consisted of a balanced detector and local oscillator (LO) laser. Here, we assumed ideal phase-matching between the LD and LO for the sake of simplicity.

Table 1. Fiber Parameters.

Parameter		SMF	DCF	Unit
Loss coefficient	α	0.2	0.45	dB/km
Dispersion coefficient	D	+17.0	−80.0	ps/nm/km
Dispersion slope	S	0.057	−0.22	ps/nm^2/km
Effective core cross-section	A_{eff}	80	14	μm^2
Nonlinear index coefficient	n_2	2.9×10^{-20}	4.3×10^{-20}	m^2/W

As our primary intention is to evaluate the reduction of the SPM effect during transmission, we neglect all the electrical and optical noises in the system. Since PAPR is closely related to waveform degradation, eye diagrams of the received signal were used to evaluate the transmitted signal. OSSB-SC signal eye diagrams of before and after transmission at a modulation depth of 0.1 are compared in Figure 14. Figure 14a depicts the back-to-back eye diagram, and the eye diagram of the received signal is presented in Figure 14b. We define the parameter k as the ratio of LO power to received signal power. k was set to 20 dB and P_{IN} to 9 dBm for the calculations of transmitted signal eye diagrams. Very obviously, high PAPR of OSSB-SC degrades the received signal eye diagrams due to the nonlinear phase-shifts by SPM. Figure 15 compares the eye diagrams of the transmitted signal whose PAPR is reduced using the techniques introduced in Sections 3–5.

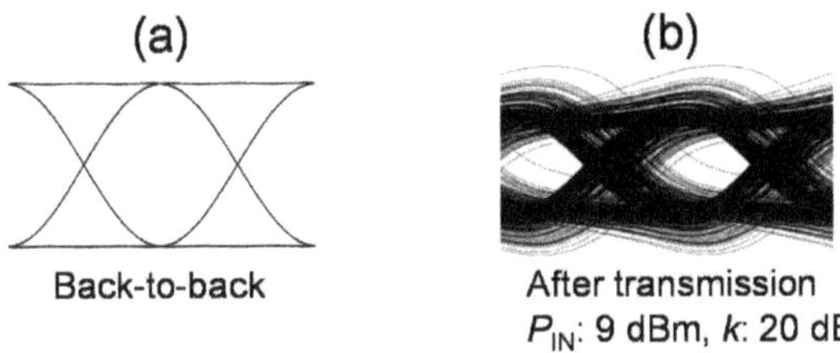

Figure 14. Back-to-back and transmitted signal eye diagrams, (**a**) back-to-back, (**b**) after 100-km transmission.

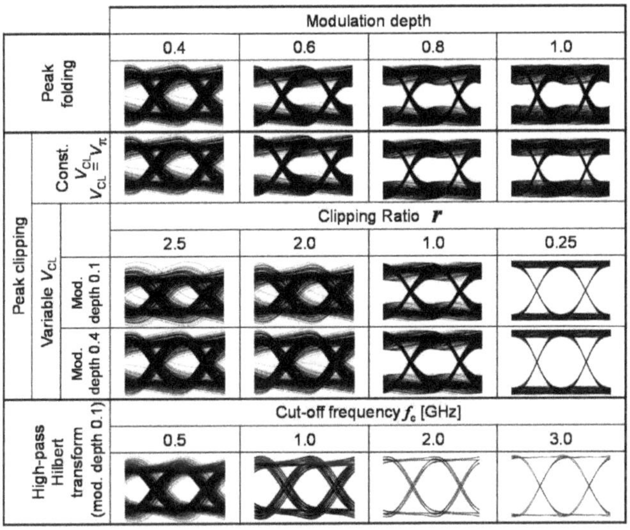

		Modulation depth			
		0.4	0.6	0.8	1.0

Figure 15. Transmitted signal eye diagrams of PAPR reduced OSSB-SC signal (P_{IN} 9 dBm, k 20 dB).

Eye diagrams of the peak-folded OSSB-SC signal are shown at the top of Figure 15 when modulation depth is varied. Eye-opening becomes larger with increasing modulation depth in peak-folded OSSB-SC signal. Because the peak-to-peak voltage of Hilbert-transformed signal becomes larger than $2V_\pi$ for modulation depths greater than 0.4, peak-folding of Hilbert-transformed signal component occurs which effectively increases the eye-opening.

The next row of Figure 15 displays the transmitted signal eye diagrams of peak-clipped OSSB-SC signal. V_{CL} was set to be V_π and modulation depth was varied. Compared with peak-folding, the eye-opening of the peak-clipped signal slightly increase. The reason for this slight increase of eye-opening is the reduction of high order harmonics during the modulation, by peak clipping.

Peak-clipped signal eye diagrams with variable clipping voltage are also shown. When r is large, eye diagrams of modulation depth 0.1 show less eye-opening than that of modulation depth 0.4. This is because of the nonlinearity of the modulator. For small modulation depths, the modulator operates in its linear region. Hence, the PAPR of the OSSB-SC signal becomes high due to the linearly transformed peaks of Hilbert-transformed from the electrical domain to the optical domain. For larger modulation depths, PAPR becomes less since modulator nonlinearity folds the peaks of Hilbert-transformed signal. For smaller values of r, the clipping amount of the Hilbert-transformed waveform increases. As a result, the baseband signal component becomes dominant. The effect of modulation depth on the transmitted signal eye degradations disappears giving similar eye openings in both modulation depths.

The most opened eyes are archived in high-pass Hilbert-transformed OSSB-SC signal. Despite higher PAPR, high-pass Hilbert transform gives less degraded eyes compared to peak-folding and peak-clipping methods. This can be found comparing the eye diagrams of Figure 15. One can compare the eye diagrams of peak-folded and peak-clipped ($V_{CL} = V_\pi$) signal of modulation depths of 0.4 (PAPR = 2.51) and 0.6 (PAPR = 1.86) with the eye diagrams of high-pass Hilbert transform at cut-off frequencies of 0.5 GHz (PAPR = 3.05) and 1.0 GHz (PAPR = 2.18). Degradations of high-pass Hilbert-transformed OSSB-SC signal become less due to the deficiency of high order harmonics in the modulated signal spectrum. Since there are no waveform degradations of Hilbert-transformed signal during the PAPR reduction process, broadened mark, space levels are seen in peak-folding and peak-clipping do not appear in high-pass Hilbert-transformed eye diagrams. Along with increasing cut-off frequency of the high-pass Hilbert transformer, eye-opening increases due to the reduction of PAPR.

To evaluate the improvement of eye-opening by proposed techniques, eye-opening penalty (*EOP*) is defined as follows

$$EOP = \frac{E_R}{(\alpha_{SMF}L_{SMF} \times \alpha_{DCF}L_{DCF})E_T},\tag{4}$$

where E_R is the eye-opening (*EO*) of transmitted signal and E_T is the eye-opening of the back-to-back eye diagram when modulation depth is 0.1. *EO* is defined as the ratio of ΔI_r to $(\Delta I_r)_{max}$ where ΔI_r and $(\Delta I_r)_{max}$ are inside and outside eye openings as shown in the inset of Figure 16. α and L denote loss coefficients and fiber lengths of SMF and DCF of the transmission link, respectively.

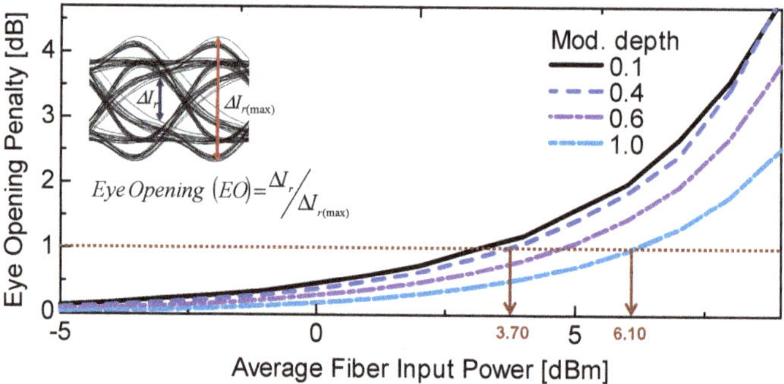

Figure 16. Eye-opening penalty of peak-folded OSSB-SC signal, inset: definition of eye-opening (*EO*), ΔI_r, and $(\Delta I_r)_{max}$ are inside and outside eye openings, respectively.

Figures 16–18 show the *EOP* of the three methods introduced in Sections 3–5, respectively. In each case, *EOP* increases exponentially with P_{IN}. This increase in *EOP* is due to the waveform degradations caused by SPM. As noted with the eye diagrams, the increase in *EOP* was alleviated at higher modulation depths for peak folded and peak clipped OSSB-SC signal. This is because of the peak reduction of the Hilbert-transformed signal component by peak folding and peak clipping at greater modulation depths, respectively. In the high-pass Hilbert transform method, the average fiber input power which *EOP* starts to increase exponentially becomes higher with increasing cut-off frequency.

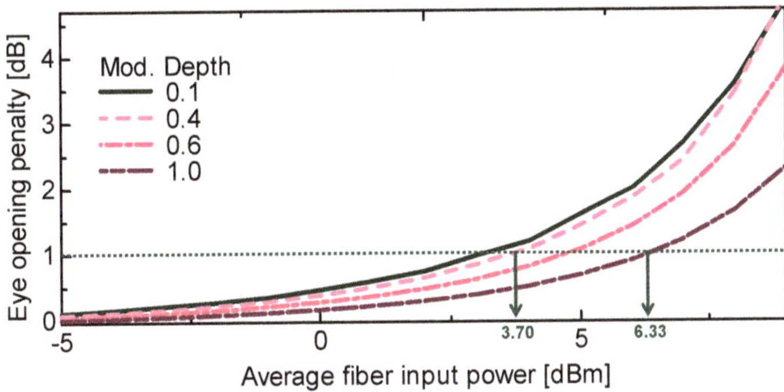

Figure 17. Eye-opening penalty of peak-clipped OSSB-SC signal ($V_{CL} = V_\pi$).

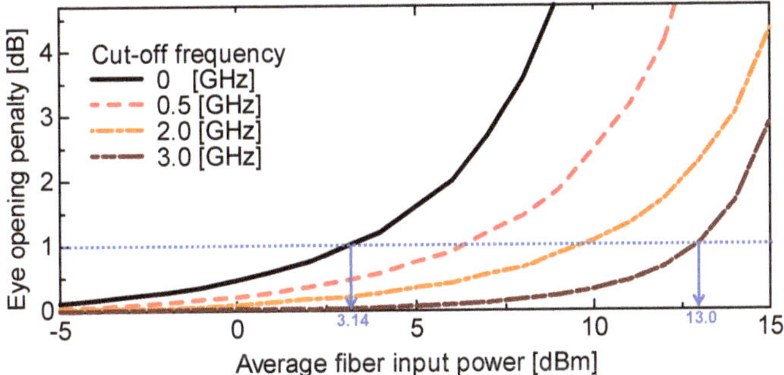

Figure 18. Eye-opening penalty of high-pass Hilbert-transformed OSSB-SC signal.

We defined the value of the average fiber input power that *EOP* becomes 1 dB as the SPM threshold. We found the SPM threshold of our system for the OSSB-SC transmission as 3.14 dBm when the modulation depth is 0.1. Peak-folding starts to occur for modulation depths larger than 0.4. SPM threshold becomes 3.99 dBm at the modulation depth of 0.4 where peak-folding just started. In the case of peak clipping, peak clipping also starts to occur for modulation depths larger than 0.4 since $V_{CL} = V_\pi$. The SPM threshold of the peak-clipped OSSB-SC signal at a modulation depth of 0.4 is 3.70 dBm. In peak folding and peak clipping methods, the SPM threshold can be improved up to 2.40 dB and 2.63 dB respectively, comparing to the situation where peak folding or clipping is not used. In the high-pass Hilbert transformed method, an SPM threshold of 9.86 dB is achieved at the cut-off frequency of 3 GHz for 10 Gbit/s NRZ-coded signals.

7. Discussion

Even though OSSB transmission is attracting the interest of researchers' as a cost-effective solution for short-reach links, high PAPR of OSSB-SC transmission becomes a major drawback during the transmission. PAPR reduction has been substantially studied in wireless transmission. However, PAPR reduction methods for optical links have to be investigated.

We previously proposed three techniques to reduce the PAPR of optical SSB-SC signal using both time and frequency-domain signal processing. Peak folding using the nonlinearity of optical modulator was originally studied to suppress the noise of driving signals [38]. Later, we reported the capability of PAPR reduction of OSSB-SC signal using the same property of the LN modulator. In the peak-folding method, peaks of the Hilbert-transformed signal are folded back using the sinusoidal transfer function of the LN modulator. The Peak power was approximately halved while maintaining 20-dB spectral suppression. An SPM threshold improvement of 2.40 dB is reported.

PAPR reduction by peak clipping has been studied extensively for OFDM transmission in wireless communication [32,33]. We investigated PAPR reduction of OSSB-SC signal by clipping peaks of the Hilbert-transformed signal in the electrical domain. Reduction in spectral suppression is noticed during PAPR reduction. Peak power of the optical SSB-SC signal was reduced to about one-half of the original value with a spectral suppression of 20 dB. Using peak-clipping, the SPM threshold is improved by 2.63 dB.

To the best of our knowledge for the first time, we proposed a PAPR reduction method for OSSB-SC transmission using frequency-domain signal processing. Peaks of Hilbert-transformed signal are suppressed by reducing the power of low-frequency components of the spectrum where the energy is concentrated. The all-pass amplitude response of the Hilbert transformer was modified to a high-pass response to reduce the power of lower frequency components of the spectrum. The high-pass Hilbert transform method reduces the PAPR of the optical SSB-SC signal from 4.17 to 1.65. Bandwidth saving of over 30% was achieved relative to double-sideband modulation. As a result of PAPR reduction, the SPM threshold was improved by 9.86 dB.

In this study, we chose 10G-class BPSK modulation for the sake of simplicity. However, the proposed concepts can be extended to other modulation formats and higher transmission rates.

8. Conclusions

We compared the performances of three previously proposed methods to reduce the PAPR of OSSB-SC signal in a repeater-less 100-km transmission link. The effectiveness of the proposed methods was confirmed by the analysis of the transmitted signal. SPM threshold of the studied system can be improved by 2.40 dB and 2.63 by peak folding and peak clipping of Hilbert-transformed signal respectively. Besides reducing PAPR, the peak-folding method brings the benefit of driving signal noise suppression. (In this study, we focused on a noise-free signal for the sake of simplicity). Among the proposed methods, the high-pass Hilbert-transform method makes OSSB-SC signal most tolerant to SPM-based signal degradations. SPM threshold can be adjusted according to the demand by choosing the appropriate cut-off frequency in the high-pass Hilbert transform method. For 10 Gbit/s NRZ-coded baseband signal, the SPM threshold of 13 dBm could be achieved by setting the cut-off frequency to 3 GHz.

It is seen that the reduction of spectral efficiency cannot be avoided during PAPR reduction. Almost similar spectral characteristics were noticed in peak clipping and peak folding methods. Different from the other two methods, spectral bandwidth increase is noticed in the high-pass Hilbert-transform method during PAPR reduction. Hence, a technique to reduce PAPR should be chosen after taking the available bandwidth into account.

Chromatic dispersion of optical fibers has been identified as a limiting factor of next-generation radio over fiber (RoF) systems [39]. The reach limitation caused by the fiber dispersion is predicted to be severe in radio access networks where capacity improvements are planned to achieve using higher frequencies such as millimeter waves. Different methods such as optical-domain and electrical-domain compensation and O-band transmission have been studied to circumvent this issue of chromatic dispersion [40]. Among those, SSB transmission becomes a strong candidate because of the colorless operation capability and the simple configuration of the receiver. In combination with the proposed PAPR reduction methods, OSSB-SC transmission can increase the transmission power efficiency other than extending the reach.

Author Contributions: Conceptualization, K.I.A.S. and K.T.; investigation, K.I.A.S., K.T.; writing—original draft preparation, K.I.A.S.; resources and discussion, J.M.; K.T., review and editing, K.T., J.M. All authors have read and agreed to the published version of the manuscript.

Funding: This work was partly supported by JSPS KAKENHI Grant Numbers 19H02134, 20K14739.

References

1. *Technology Options for 400G Implementation. OIF-Tech- Options-400G-01.0*; Optical Internetworking Forum (OIF): Fremont, CA, USA, 2015.
2. The 2018 Ethernet Roadmap. Available online: https://ethernetalliance.org/the-2018-ethernet-roadmap/ (accessed on 28 December 2020).
3. Bakopoulos, P.; Ma, P.; Tsiokos, D.; Hettrich, H.; Eltes, F.; Uhl, C.; Lischke, S.; Petousi, D.; Abel, S.; Schmid, R.; et al. Scaling Optical Interconnects Beyond 400 Gb/s. In Proceedings of the 2018 European Conference on Optical Communication (ECOC), Rome, Italy, 23–27 September 2018; pp. 1–3.
4. Hu, Q.; Che, D.; Wang, Y.; Shieh, W. Advanced Modulation Formats for High-performance Short-reach Optical Inter-connects. *Opt. Exp.* **2015**, *23*, 3245–3259. [CrossRef] [PubMed]
5. Cartledge, J.C.; Karar, A.S. 100 Gb/s Intensity Modulation and Direct Detection. *J. Light. Technol.* **2014**, *32*, 2809–2814. [CrossRef]
6. Shi, J.; Zhang, J.; Zhou, Y.; Wang, Y.; Chi, N.; Yu, J. Transmission Performance Comparison for 100-Gb/s PAM-4, CAP-16, and DFT-S OFDM With Direct Detection. *J. Light. Technol.* **2017**, *35*, 5127–5133. [CrossRef]
7. Zhou, X.; Nelson, L. Advanced DSP for 400 Gb/s and Beyond Optical Networks. *J. Light. Technol.* **2014**, *32*, 2716–2725. [CrossRef]
8. *Flex Coherent DWDM Transmission Framework Document. OIF-FD-FLEXCOH-DWDM-01.0*; Optical Internetworking Forum (OIF): Fremont, CA, USA, 2017.
9. Weber, M. Towards an objective for 400 Gb/s for DCI applications. In Proceedings of the IEEE802 Study Group Ad hoc Meeting, Chicago, IL, USA, 8 May 2018.
10. Kikuchi, N.; Hirai, R.; Fukui, T. Quasi Single-Sideband (SSB) IM/DD Nyquist PAM Signaling for High-Spectral Efficiency DWDM Transmission. In Proceedings of the Optical Fiber Communication Conference, Anaheim, CA, USA, 20–22 March 2016; The Optical Society: Washington, DC, USA, 2016.
11. Zhang, L.; Zuo, T.; Mao, Y.; Zhou, E.; Liu, G.N.; Xu, X. Beyond 100-Gb/s Transmission Over 80-km SMF Using Direct-Detection SSB-DMT at C-Band. *J. Light. Technol.* **2016**, *34*, 723–729. [CrossRef]
12. Zhu, Y.; Ruan, X.; Zou, K.; Zhang, F. Beyond 200G Direct Detection Transmission with Nyquist Asymmetric Twin-SSB Signal at C-Band. *J. Light. Technol.* **2017**, *35*, 3629–3636. [CrossRef]
13. Wan, Z.; Li, J.; Shu, L.; Fu, S.; Fan, Y.; Yin, F.; Zhou, Y.; Dai, Y.; Xu, K. 64-Gb/s SSB-PAM4 Transmission Over 120-km Dis-persion-Uncompensated SSMF With Blind Nonlinear Equalization, Adaptive Noise-Whitening Postfilter and MLSD. *J. Light. Technol.* **2017**, *35*, 5193–5200. [CrossRef]
14. Kaneda, N.; Lee, J.; Chen, Y.-K. Nonlinear Equalizer for 112-Gb/s SSB-PAM4 in 80-km Dispersion Uncompensated Link. In Proceedings of the Optical Fiber Communication Conference, Los Angeles, CA, USA, 19–23 March 2017; The Optical Society: Washington, DC, USA, 2017.
15. Schmidt, B.J.C.; Lowery, A.J.; Armstrong, J. Experimental Demonstrations of Electronic Dispersion Compensation for Long-Haul Transmission Using Direct-Detection Optical OFDM. *J. Light. Technol.* **2008**, *26*, 196–203. [CrossRef]
16. Zhong, K.; Zhou, X.; Huo, J.; Yu, C.; Lu, C.; Lau, A.P.T. Digital Signal Processing for Short-Reach Optical Communications: A Review of Current Technologies and Future Trends. *J. Light. Technol.* **2018**, *36*, 377–400. [CrossRef]
17. Peng, W.-R.; Wu, X.; Feng, K.-M.; Arbab, V.R.; Shamee, B.; Yang, J.-Y.; Christen, L.C.; Willner, A.E.; Chi, S. Spectrally efficient direct-detected OFDM transmission employing an iterative estimation and cancellation technique. *Opt. Exp.* **2015**, *17*, 9099–9111. [CrossRef] [PubMed]
18. Randel, S.; Pilori, D.; Chandrasekhar, S.; Raybon, G.; Winzer, P. 100-Gb/s discrete-multitone transmission over 80-km SSMF using single-sideband modulation with novel interference-cancellation scheme. In Proceedings of the 2015 European Conference on Optical Communication (ECOC), Valencia, Spain, 27 September–1 October 2015; Institute of Electrical and Electronics Engineers (IEEE): New York, NY, USA; pp. 1–3.
19. Ju, C.; Liu, N.; Chen, X.; Zhang, Z. SSBI Mitigation in A-RF-Tone-Based VSSB-OFDM System with a Frequency-Domain Volterra Series Equalizer. *J. Light. Technol.* **2015**, *33*, 4997–5006. [CrossRef]
20. Li, Z.; Erkilinc, M.S.; Shi, K.; Sillekens, E.; Galdino, L.; Xu, T.; Thomsen, B.C.; Bayvel, P.; Killey, R.I. Digital Linearization of Direct-Detection Transceivers for Spectrally Efficient 100 Gb/s/λ WDM Metro Networking. *J. Light. Technol.* **2018**, *36*, 27–36. [CrossRef]
21. Li, Z.; Erkilinc, M.; Shi, K.; Sillekens, E.; Galdino, L.; Thomsen, B.; Bayvel, P.; Killey, R. Joint Optimisation of Resampling Rate and Carrier-to-Signal Power Ratio in Direct-Detection Kramers-Kronig Receivers. In Proceedings of the 2017 European Conference on Optical Communication (ECOC), Gothenburg, Sweden, 17–21 September 2017; pp. 1–3.
22. Fujita, T.; Toba, K.; Sampath, K.I.A.; Maeda, J. Impact of Sampling and Quantization on Kramers-Kronig Relation-Based Direct Detection. *IEICE Trans. Commun.* **2020**, 1291–1298. [CrossRef]
23. Li, Z.; Erkilinc, M.S.; Shi, K.; Sillekens, E.; Galdino, L.; Thomsen, B.C.; Bayvel, P.; Killey, R.I.; Erkılınç, M.S. 168 Gb/s/λ Direct-Detection 64-QAM SSB Nyquist-SCM Transmission over 80 km Uncompensated SSMF at 4.54 b/s/Hz net ISD using a Kramers-Kronig Receiver. In Proceedings of the 2017 European Conference on Optical Communication (ECOC), Gothenburg, Sweden, 17–21 September 2017; pp. 1–3.
24. Bo, T.; Kim, H. Kramers-Kronig receiver operable without digital upsampling. *Opt. Express* **2018**, *26*, 13810–13818. [CrossRef]
25. Le, S.T.; Schuh, K.; Chagnon, M.; Buchali, F.; Dischler, R.; Aref, V.; Buelow, H.; Engenhardt, K.M. 1.72-Tb/s Virtu-al-Carrier-Assisted Direct-Detection Transmission Over 200 km. *J. Light. Technol.* **2017**, *36*, 1347–1353. [CrossRef]

26. Chen, X.; Antonelli, C.; Chandrasekhar, S.; Raybon, G.; Mecozzi, A.; Shtaif, M.; Winzer, P. 4 × 240 Gb/s Dense WDM and PDM Kramers-Kronig Detection with 125-km SSMF Transmission. In Proceedings of the 2017 European Conference on Optical Communication (ECOC), Gothenburg, Sweden, 17–21 September 2017; pp. 1–3.

27. Takano, K.; Murakami, T.; Sawaguchi, Y.; Nakagawa, K. Influence of self-phase modulation effect on waveform degradation and spectral broadening in optical BPSK-SSB fiber transmission. *Opt. Exp.* **2011**, *19*, 9699–9707. [CrossRef] [PubMed]

28. Sampath, K.I.A.; Takano, K. PAPR reduction technique for optical SSB modulation using peak folding. In Proceedings of the 2015 Opto-Electronics and Communications Conference (OECC), Shanghai, China, 28 June–2 July 2015; pp. 1–3.

29. Sampath, K.I.A.; Takano, K. Performance evaluation of peak-clipped optical BPSK-SSB modulated signal. *Opt. Express* **2015**, *23*, 19813–19820. [CrossRef]

30. Sampath, K.I.A.; Takano, K. Phase-Shift Method-Based Optical VSB Modulation Using High-Pass Hilbert Transform. *IEEE Photon J.* **2016**, *8*, 1–13. [CrossRef]

31. Lathi, B.P.; Ding, Z. *Modern Digital and Analog Communication Systems*; Oxford University Press: New York, NY, USA, 2009.

32. Li, X.; Cimini, L. Effects of clipping and filtering on the performance of OFDM. *IEEE Commun. Lett.* **1998**, *2*, 131–133.

33. Lim, D.-W.; Heo, S.-J.; No, J.-S. An overview of peak-to-average power ratio reduction schemes for OFDM signals. *J. Commun. Networks* **2009**, *11*, 229–239. [CrossRef]

34. Sampath, K.I.A.; Takano, K.; Sato, M. Transmission performance of peak-clipped carrier-suppressed optical SSB signal with variable clipping voltage. In Proceedings of the 2015 IEEE 10th International Conference on Industrial and Information Systems (ICIIS), Peradeniya, Sri Lanka, 18–20 December 2015; pp. 535–539.

35. Zhang, B.; Malouin, C.; Schmidt, T.J. Design of coherent receiver optical front end for unamplified applications. *Opt. Express* **2012**, *20*, 3225–3234. [CrossRef] [PubMed]

36. Januario, J.C.S.S.; Chiuchiarelli, A.; Rossi, S.M.; Cruz, J.H.; Reis, J.D.; Mornatta, C.; Festa, A.; Bordonalli, A.C.; Junior, J.H.C. System Design for High-Capacity Unrepeated Optical Transmission. *J. Light. Technol.* **2019**, *37*, 1246–1253. [CrossRef]

37. Agrawal, G.P. *Nonlinear Fiber Optics*; Academic: New York, NY, USA, 2010.

38. Ip, E.; Kahn, J. Carrier synchronization for 3- and 4-bit-per-symbol optical transmission. *J. Light. Technol.* **2005**, *23*, 4110–4124. [CrossRef]

39. *Radio over Fiber Systems, Recommendation ITU-T G.9803 Amendment 1*; International Telecommunication Union: Geneva, Switzerland, 2019.

40. Zou, J.; Sasu, S.A.; Lawin, M.; Dochhan, A.; Eiselt, J.-P.E.M. Advanced optical access technologies for next-generation (5G) mobile networks [Invited]. *J. Opt. Com. Netw.* **2020**, *12*, D86–D98. [CrossRef]

Communication

Performance Enhancement of DWDM-FSO Optical Fiber Communication Systems Based on Hybrid Modulation Techniques under Atmospheric Turbulence Channel

Mohammed R. Hayal [1,*], **Bedir B. Yousif [2,3]** and **Mohamed A. Azim [1]**

[1] Electronics and Communications Engineering Department, Faculty of Engineering, Mansoura University, Mansoura 35516, El-Dakahlia Governorate, Egypt; mazim12@mans.edu.eg

[2] Electrical Engineering Department, Electronics and Communications Engineering Branch, Faculty of Engineering, Kafrelsheikh University, Kafrelsheikh 33511, Kafrelsheikh Governorate, Egypt; bedir.yousif@eng.kfs.edu.eg

[3] Electrical Engineering Department, Faculty of Engineering and Information Technology-Onaizah Colleges, Ash Shabili 4403, Al-Qassim 56447, Saudi Arabia

* Correspondence: mohammedraisan@gmail.com

Citation: Hayal, M.R.; Yousif, B.B.; Azim, M.A. Performance Enhancement of DWDM-FSO Optical Fiber Communication Systems Based on Hybrid Modulation Techniques under Atmospheric Turbulence Channel. *Photonics* **2021**, *8*, 464. https://doi.org/10.3390/photonics8110464

Received: 26 August 2021
Accepted: 20 October 2021
Published: 22 October 2021

Publisher's Note: MDPI stays neutral with regard to jurisdictional claims in published maps and institutional affiliations.

Abstract: In this paper, we enhance the performance efficiency of the free-space optical (FSO) communication link using the hybrid on-off keying (OOK) modulation, M-ary digital pulse position modulation (M-ary DPPM), and M-pulse amplitude and position modulation (M-PAPM). This work analyzes and enhances the bit error rate (BER) performance of the moment generating function, modified Chernoff bound, and Gaussian approximation techniques. In the existence of both an amplified spontaneous emission (ASE) noise, atmospheric turbulence (AT) channels, and interchannel crosstalk (ICC), we propose a system model of the passive optical network (PON) wavelength division multiplexing (WDM) technique for a dense WDM (DWDM) based on the hybrid fiber FSO (HFFSO) link. We use eight wavelength channels that have been transmitted at a data rate of 2.5 Gbps over a turbulent HFFSO-DWDM system and PON-FSO optical fiber start from 1550 nm channel spacing in the C-band of 100 GHz. The results demonstrate (2.5 Gbps × 8 channels) 20 Gbit/s-4000 m transmission with favorable performance. In this design, M-ary DPPM-M-PAPM modulation is used to provide extra information bits to increase performance. We also propose to incorporate adaptive optics to mitigate the AT effect and improve the modulation efficiency. We investigate the impact of the turbulence effect on the proposed system performance based on OOK-M-ary PAPM-DPPM modulation as a function of M-ary DPPM-PAPM and other atmospheric parameters. The proposed M-ary hybrid DPPM-M-PAPM solution increases the receiver sensitivity compared to OOK, improves the reliability and achieves a lower power penalty of 0.2–3.0 dB at low coding level (M) 2 in the WDM-FSO systems for the weak turbulence. The OOK/M-ary hybrid DPPM-M-PAPM provides an optical signal-to-noise ratio of about 4–8 dB of the DWDM-HFFSO link for the strong turbulence at a target BER of 10^{-12}. The numerical results indicate that the proposed design can be enhanced with the hybrid OOK/M-DPPM and M-PAPM for DWDM-HFFSO systems. The calculation results show that PAPM-DPPM has increased about 10–11 dB at BER of 10^{-12} more than the OOK-NRZ approach. The simulation results show that the proposed hybrid optical modulation technique can be used in the DWDM-FSO hybrid links for optical-wireless and fiber-optic communication systems, significantly increasing their efficiency. Finally, the use of the hybrid OOK/M-ary DPPM-M-PAPM modulation schemes is a new technique to reduce the AT, ICC, ASE noise for the DWDM-FSO optical fiber communication systems.

Keywords: ASE noise; ICC; DWDM-FSO/PON optical fiber network; hybrid OOK/M-ary DPPM-M-PAPM; hybrid fiber FSO (HFFSO) link

1. Introduction

Pulse position modulation (PPM) and digital PPM (DPPM) and systems are modulation schemes that can perform great performance in free-space optical (FSO) transmission links [1,2]. This format is used in a variety of applications, including FSO links, hybrid fibers, optical wireless communication (OWC), subsequent FSO systems, satellite-to-satellite systems, atmospheric turbulence (AT), interchannel crosstalk (ICC), and indoor wireless channels [1–3]. The modulation of DPPM improves power efficiency and does not require monitoring of decision-making circuit thresholds [4,5]. Many experiments on the hybrid fiber/FSO (HFFSO) systems and the OWC have been carried out [5–9]. Previous researches have demonstrated that the DPPM and PPM schemes outperform on-off keying (OOK) in terms of sensitivity and power efficiency for the HFFSO link. The M-pulse amplitude and position modulation (M-PAPM) for both pulse amplitude modulation (PAM) and PPM modulation have been investigated and studied for optical fiber (OF) communications [5–8]. M-PAPM can provide high efficiency and sensitivity in the FSO communication; because the dispersion is free [1,5,8]. Because of the increased bandwidth requirement for higher data rates, OF, AT, OWC and indoor networks [8–15] have been proposed to multiplex the wavelength division multiplexing (WDM) technique and dense WDM (DWDM) systems. The WDM technique could also be used in the HFFSO, OWC, and hybrid OF/multiple networks. For example, the proposed approach designs a high-performance system and bandwidth optimization solution with long-range potential, higher bit rates, high-speed technology, and enhanced data protection for WDM passive optical networks (WDM-PON) [8,9,13,16]. The WDM application has been presented in both OF and FSO systems [8,9,17–21]. A suitable technique for representing amplified spontaneous emission noise (ASE) in an FSO transmission link is the moment generating function (MGF), although we have upper limits at the bit-error-rate (BER) using the updated techniques of Chernoff bound (CB), Gaussian approximating (GA), and modified Chernoff bound (MCB) [5,8,9,22–25]. Both the FSO and OF systems have been presented using WDM technology [8,9,17–21]. The moment generating function (MGF) represents a suitable technique for finding the amplified spontaneous emission (ASE) noise in an FSO transmission link [2,9,17]. Using the modified Chernoff bound (MCB) technique, Gaussian approximations (GA), and Chernoff bound (CB), we obtain upper bounds on the bit error rate (BER) [5,8,9,21–25]. To reduce the AT impact, approaches are proposed including compensation for the mitigation techniques and adaptive optics (AO) focused on digital signal processing [22]. We highlight the main contributions, including: (1) we use the M-ary DPPM-M-PAPM modulation to deliver additional bits to improve the efficiency; (2) we add the AO to minimize the ICC interferences and enhance the reliability performances; (3) we achieve adequate BER results with a lower complexity; (4) we extract the theoretical BER expressions and provide simulation results for M-DPPM-M-PAPM modulation schemes in the WDM-PON/HFFSO scenarios; (5) the hybrid OOK/M-ary DPPM-M-n-PAPM and the AO are offered to boost the performance and system efficiency in receivers of WDM-HFFSO systems through the OOK non-return-to-zero (OOK-NRZ) modulation; and (6) we improve the power penalty (PP) performance and the system efficiency for the WDM-HFFSO systems, optical-wireless and fiber-optic communication systems. In this work, we enhance the hybrid OOK/DPPM-M-PAPM techniques and improve the signal-to-noise-ratios (SNRs) of the HFFSO systems under the AT effects, ICC, and ASE. In this work, we enhance the hybrid OOK/DPPM-M-PAPM systems and improve the signal-to-noise-ratios (SNRs) of the HFFSO under the AT, ICC, and ASE effects. We develop the proposed model in the [9,17]. Also, we enhance our calculations and evaluate them to reduce the AT, and ASE noise. The remainder of the paper is organized as follows: Section 2 describes the proposed PON/WDM-HFFSO optical fiber communication system. Section 3 discusses the M-ary DPPM-M-PAPM model for the hybrid WDM-PON/HFFSO link. The AT channel effects are analyzed in Section 4. The numerical results are presented in Section 5. Section 6 concludes this paper.

2. System Description

The length of slots $t_s = MT_b/n$ is for the DPPM frames, $n = 2^M$ where $T_b = 1/R_b$ is for the bit cycle, R_b for the data rate and M is for the coding level (CL) [9]. Figure 1a shows the HFFSO-PON systems where optical signals (OSs) suffer from the ASE noise (ASEN), beam-spreading, beam-absorption, attenuation, ICC, and splitting losses at the optical band-pass filter (OBPF)/demux. The proposed system is modeled and simulated using MATLAB software (2013). The results are based on Monte-Carlo simulations. We use eight channels WDM-DWDM starting from 1550 nm over single-mode fiber and the channel spacing in the C-band of 100 GHz of the ITU (International Telecommunication Union). The results demonstrate (2.5 Gbps data rate $\times$8 channels) 20 Gbit/s-4000 m transmission with favorable performance. We investigate eight-wavelength channels that have been transmitted over turbulent HFFSO links using the WDM technique. The feeder fiber path length of 20 km is proposed for the system model [8,15,17]. By using the optical amplifier (OA) gain G on the remote node, we improve the device model after receiver collection lens (RCL) as seen in Figure 1a. We propose that the OA system is used to reduce the inter-channel crosstalk for the proposed system. Figure 1b shows the proposed architecture for all scenarios [9]. As seen in Figure 1, M-ary DPPM is transmitted over the OA and OBPF (b). The receiver converts the OS into an electrical signal. The module of the receivers consists of a photodetector (PD), an electric amplifier, a filter, and a compared circuitry for the decision circuit and AO. We propose the hybrid pulse modulation/M-ary DPPM-M-PAPM for the DWDM-FSO and hybrid optical fiber over the atmospheric turbulence channel. Information is transmitted as a series of DPPM-M-PAPM pulses. The modulated signal from M-ary DPPM-M-PAPM is connected to OA and OBPF to provide M-ary DPPM-M-PAPM output as shown in Figure 1b [9]. For our calculations, we assume the loss of the signal multiplexer (mux)/demultiplexer (demux) ($\leq$3.5 dB) [8,15].

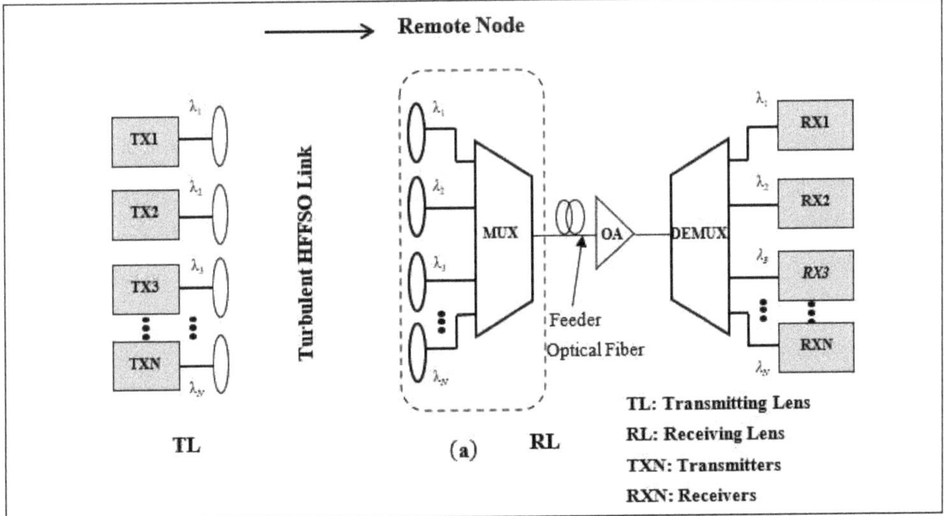

Figure 1. *Cont.*

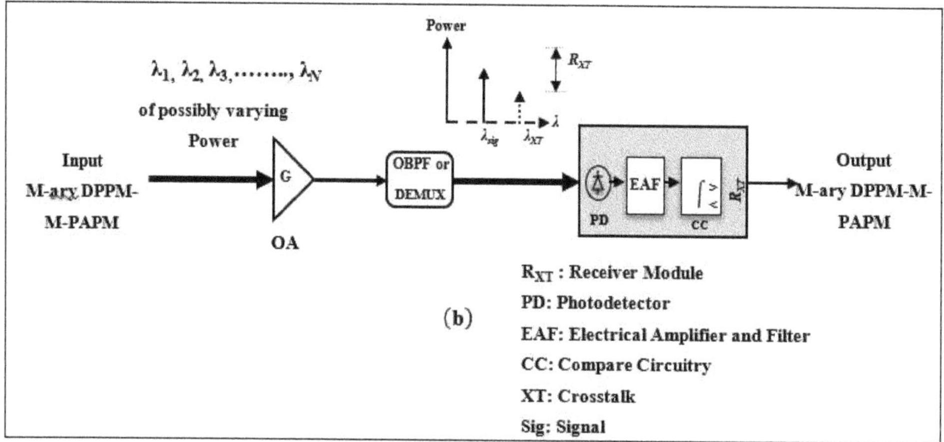

Figure 1. System model for 8 channels PON-DWDM-HFFSO for OOK-NRZ/M-ary DPPM-M-PAPM under moderate turbulence (MT), strong turbulence (ST), and weak turbulence (WT): (a) proposed framework for interchannel crosstalk (ICC) evaluation and (b) schematic of the receiver system. We reproduced it from [9].

3. M-ary Digital Pulse Position Modulation and M-pulse Amplitude and Position Modulation (M-ary DPPM-M-PAPM) Scheme

In this section, the random variable (RV) of the MGF describes the current $Y_{sig}(\Delta t)$ where sig $\in \{0, 1\}$ which depends on the pulses or is not transmitted for the signal pulses, Δt is the duration of the crosstalk pulse. It is written as [5,9,24,26–28]:

$$M_{Y_{sig}(\Delta t)}(s) = \left\{ \frac{R'G\left(e^{sq/t_s} - 1\right) \int_{t_s} sigP_{tr}(t)dt}{1 - R'N_0\left(e^{sq/t} - 1\right)} \right\} \frac{\exp\left\{ \frac{R_i'G\left(e^{sq/t_s} - 1\right) \int_{\Delta t} P_{XT}(t)dt}{1 - R_i'N_{o\_XT}\left(e^{sq/t_s}\right)} \right\}}{\left[1 - R'N_0\left(e^{sq/t_s} - 1\right)\right]^L} \tag{1}$$

where $\Delta t = t_s$ is the time slots align with the OS slots otherwise t_1 or t_2, and $\Delta t = 0$ for no crosstalk in the slot. Furthermore, the DPPM-PAPM pulse and ICC pulse strength are P_{tr}, P_{XT}, respectively for the hybrid modulation techniques over the FSO link. $R' = \eta/hv_i, \eta$ is the PD quantum efficiency, v_i, v and are the optical frequencies of ICC wavelengths and signal respectively and h is Planck's constant, q is the electron charge, [5,8–10,29–34], $N_0 = 0.5(NF \times G - 1)\, hv$ is the OA power spectral density (PSD) at the single polarization ASEN. NF and G are the noise figure and OA gain G respectively, $L = B_o m_t t_s$ encompass the system modes for the spatial and temporal method [2,8,9,35–42], B_0 is the bandwidth of the optical noise for the demux channel and m_t is the number of ASEN states. N_{0_XT} is the PSD-ASEN at the PD and the signal-to-crosstalk ratio $C_{XT} = P_{tr}/P_{XT}$. The total MGF for Gaussian zero-mean, including the thermal noise variance (TNV) is calculated as [5,9,35–42]:

$$M_{Y_{sig}(\Delta t)}(s) = M_{Y_{sig}(\Delta t)}(s) \exp\left(\frac{s^2 \sigma_{th-DPPM-PAPM}^2}{2} \right) \tag{2}$$

where $\sigma_{th-DPPM}^2$ is the DPPM-PAPM TNV. The means and variances are given as [5,9]:

$$\mu_{Y_{sig}(\Delta t)}(s) = \frac{LR'qN_o}{t_s} + R'Gq\left(sig\, P_{tr} \frac{P_{XT}\Delta t}{t_s} \right) \tag{3}$$

$$\sigma_{X_{sig}(\Delta t)}^2 = \sigma_{th-DPPM-PAPM}^2 + \left(\frac{LR'q^2N_0(1 + R'N_0)}{t_s^2} \right) + R'Gq^2\left[(1 + 2R'N_0)\frac{sigP_{tr}}{t_s} \right] + R_i'Gq^2\left[(1 + 2R_i'N_{o_{XT}})\frac{P_{XT}\Delta t}{t_s^2} \right] \tag{4}$$

The symbol error probability (SER) is received in the exist of ICC $P_{ws(I_i - r_i)}$:

$$P_{ws(I_i - r_i)} \geq \prod_{\substack{j=1 \\ j \neq \text{sigslot}}}^{n} P\left(X_1(\Delta t) > X_j\right) \tag{5}$$

where X_j denotes the non-signal slot $X_o(\Delta t_j)$ and Δt_j is the j (empty) slot overlap with the ICC. Using the GA, the expression $P\{X_o(\Delta t_j) > X_1(\Delta t)\}$ is explained from [1,2,5,9,10,29]:

$$P\{X_o(\Delta t_j) > X_1(\Delta t)\} = 0.5\,\text{erfc}\left(\frac{\mu_{X_1(\Delta t)} - \mu_{X_0(\Delta t_j)}}{\sqrt{2\left(\sigma^2_{X_1(\Delta t)} + \sigma^2_{X_0(\Delta t_j)}\right)}}\right) \tag{6}$$

In the proposed CB, we have fixed threshold T_{th} and the general form for RV (X) variable and a is $P\left(X > T_{th}\right) \leq E\left\{\exp\left[s\left(X - T_{th}\right)\right]\right\}, s > 0$:

$$P\{X_o(\Delta t_j) > X_1(\Delta t)\} \leq M_{X_1(\Delta t)}(-s)\, M_{X_0(\Delta t)}(s) \; (s > 0) \tag{7}$$

For the MCB [2,5] of two RVs for $X_o(\Delta t_j)$ and $X_1(\Delta t)$, $P(X > T_{th}) \leq M_X(s)e^{-sT_{th}}/s\sigma_{th}\sqrt{\pi}$ [5,9,10]. Modifying this inequality for the diffe rence of RVs s for $X_o(\Delta t_j)$ and $X_1(\Delta t)$ which both have the same TNV then:

$$P\{X_o(\Delta t_j) > X_1(\Delta t)\} \leq \frac{M_{X_1(\Delta t)}(-s)M_{X_0(\Delta t)}(s)}{2s\sigma_{th}\sqrt{\pi}}(s > 0) \tag{8}$$

The SER for MDPPM-MPAPM frames in the exist of ICC is written as [1,9,10]:

$$P_{we(I_s - r_s)} \leq 1 - [1 - P\{X_o(0) > X_1(\Delta t)\}]^{n-1(I_s - r_s)}[1 - P\{X_o(t_s) > X_1(\Delta t)\}]^{I_s - r_s} \tag{9}$$

where and I_s, r_s are the numbers of ICC system of duration t_s occurring in the frame and signal pulse respectively [42],

$$P_{we(I_1, I_2 - r_1, r_2)} \leq 1 - [1 - P\{X_o(0) > X_1(\Delta t)\}]^{n-1-\ddot{\chi}}[1 - P\{X_o(t_1) > X_1(\Delta t)\}]^{I_1 - r_1}[1 - P\{X_o(t_2) > X_1(\Delta t)\}]^{I_2 - r_2} \tag{10}$$

where I_1, I_2, and r_1, r_2 are the ICC duration t_1, t_2. $\ddot{\chi} = I_s - r_s$. The BER when without ICC is written as:

$$BER_{I_s}(n_1) = P_{f(I_s)}(n_1) = \frac{n}{2(n-1)} \times \left(P_{f(I_s)}(1)P_{we(I_s - 1)} + P_{s(I_s)}(0)P_{we(I_s - 0)}\right) \tag{11}$$

$$BER = \frac{1}{n}\sum_{n_1=1}^{n}\sum_{I_s=0}^{2} BER_{I_s}(n_1) \tag{12}$$

$$BER(n_1) = P_{f(0,0)}(n_1)\frac{n}{2(n-1)}P_{we(0,0\_0,0)} \tag{13}$$

while for the calculated BER, we have written as [1,2,29,35–42]:

$$BER_{I_1, I_2}(n_1) = \frac{1}{n}\sum_{t_1 = t_c}^{t_s} P_{f(I_1, I_2)}(n_1)\frac{n}{2(n-1)}\left\{\begin{array}{c}\left[P_{s(I_1, I_2)}(1,0)P_{we(I_1, I_{2_{1,0}})}\right] \\ +\left[P_{s(I_1, I_2)}(0,1)P_{we(I_1, I_{2_{0,1}})}\right]\end{array}\right\} \\ + P_{s(I_1, I_2)}(0,0)P_{we(I_1, I_{2_{0,0}})} \tag{14}$$

In the presence of ICC, the overall BER of the M-DPPM-M-PAPM is calculated from Equations (13) and (14) [1,2,29]:

$$\mathrm{BER}_{\text{M-DPPM-M-PAPM}} = \frac{1}{n} \sum_{n_1=1}^{n} \left\{ \begin{array}{c} \mathrm{BER}(n_1) + \mathrm{BER}_{0,1}(n_1) + \mathrm{BER}_{1,0}(n_1) \\ + \sum_{I_1=1}^{2} \left[\mathrm{BER}_{I_1,1}(n_1) + \mathrm{BER}_{I_1,2}(n_1) \right] \end{array} \right\} \tag{15}$$

4. Atmospheric Turbulence Channel

Differences in the air and variation between the refractive index structure (RIS) C_n^2 and Earth's surface temperature are responsible for the scintillation effects [8]. These findings are focused on the probability density function (pdf) of the Gamma-Gamma (GG) model [5,8,9,29–34].

$$p_{\mathrm{GG}}(h_Z) = \frac{2(\alpha\beta)^{(\alpha+\beta)/2}}{\Gamma(\alpha)\,\Gamma(\beta)} h_Z^{((\alpha+\beta)/2)-1} K_{\alpha-\beta}\left(2\sqrt{\alpha\beta h_Z}\right); \, h_Z > 0 \tag{16}$$

where h_Z is the total attenuations due to the signal (h_{sig}) and interfering (h_{int}) and pointing errors. α and β are the scattering methods of the influence of the large and small-eddies, respectively, and $\Gamma(\cdot)$ is the gamma function, and $K_n(\cdot)$ is the modified second kind Bessel function [8,10].

$$\alpha = \left\{ \exp\left[\frac{0.49\sigma_R^2}{\left(1 + 0.65d^2 + 1.11\sigma_R^{12/5}\right)^{7/6}} \right] - 1 \right\}^{-1} \tag{17}$$

$$\beta = \left\{ \exp\left[\frac{0.51\sigma_R^2 \left(1 + 0.69\sigma_R^{12/5}\right)^{-5/6}}{1 + 0.9d^2 + 0.62d^2\sigma_R^{12/5}} \right] - 1 \right\}^{-1} \tag{18}$$

$$\mathrm{d} = \sqrt{\mathrm{k}\mathrm{D}_{\mathrm{RX}}^2 / 4l_{fso}} \tag{19}$$

where d is the normalized RCLs [8–10,30,33] and D_{RX} is the RCL diameter. C_n^2 is the RIS constant, l_{fso} is the free-space distance, $k = 2\pi/\lambda$ is the wave-number, and λ is the wavelength [29,30,34–40]. The Rytov variance (RVAR) σ_R^2 characterize the different AT regimes over the GG pdf, if the resulting $\sigma_R^2 > 1$; we use the ST, if the RVAR $\sigma_R^2 \approx 1$, we use the MT, if RVAR $\sigma_R^2 < 1$, we apply the WT [8–10,29]; and if saturated, turbulence $\sigma_R^2 \to \infty$ are given as in [8–10,40–46]:

$$\sigma_R^2 = 1.23 C_n^2 \mathrm{k}^{7/6} l_{fso}^{11/6} \tag{20}$$

5. Results and Discussion

Table 1 shows the proposed parameters for the system model [proposed values] and values in Refs. [2,17,42] used in the system architecture. The proposed design parameters are specified in Table 1 [proposed] and [2,17,42]. We investigate the impact of the turbulence effect on the proposed system performance based on the OOK-M ary-DPPM modulation as a function of M-ary DPPM-PAPM and other atmospheric parameters. Figure 2 shows the high gain (G = 30 dB) of the AT impact in the case of BER. MCB and CB are about the same, through AT strength rises [9,10,43–46]. The curves of CB and MCB vary from GA. The high (a) (G) = 27 dB [42] and G = 30 dB [current work] with the ICC is shown in Figure 3. For M = 2 and (b) G = 8 dB, the present work shows the MCB, CB, and GA output. The MCB synchronizes with the GA on low G in Figure 3a but at high G is near the CB with as the ASEN decreases the TNV results. The GA at high G with no ICC and in the existence of ICC exceeds the CB and MCB. The CB extends the CL and bandwidth of the DPPM-PAPM receiver for noise equivalent B_e. The margin of the GA is greater than the MCB of the PAPM-M-DPPM. The MCB, CB, and GA output of a G = 8 dB with an ICC and M = 2 is shown in Figure 3b. The results are enhanced and validated with Ref. [42]. The MCB synchronizes with the GA at low G but gains considerably near the CB as the ASEN reduces the TNV impacts [43–46].

Table 1. System parameters used for calculations in [2,17,42] and [Proposed work].

Parameters	Description	Refs. [2,17]	Ref. [42]	[Proposed Work]
R_b	Data rate	2.5 Gbps	2.5 Gbps	2.5 Gbps
B_o	Demux OBPF bandwidth	80 GHz	76 GHz	80 GHz
λ_{sig}	Wavelength	1550 nm	1550 nm	1550 nm
η	Quantum efficiency	0.75	0.9	1
G	OA gain	30.6 dB or 8.8 dB	27 dB or 8 dB	30 dB or 8 dB
NF	OA noise figure	4.77 dB [2]	4.77 dB [2]	4.77 dB [2]
l_{fso}	FSO link length	1000 m and 1500 m	1000 m and 1500 m	1500 m and 4000 m
m_t	ASE noise polarization states	2	2	2

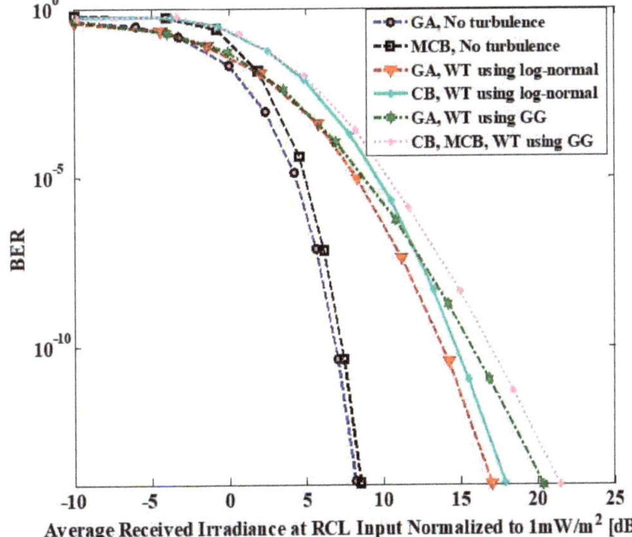

Figure 2. The bit error rate (BER) against an average irradiance [dB] for no turbulence (NT), and WT using Gamma-Gamma (GG) model and log-normal distribution.

In the case of BER targets of 10^{-12}, ST and WT using MCB, CB, and GA, the receiver sensitivity (RS) is applied for G =30.6 dB and l_{fso} = 1500 m [2,9,40] as shown in Figure 4a,b. Numerical results demonstrated that RSs are −51.49 dBm (CB), −51.59 dBm (MCB), and -50.53 dBm (GA), resulting in improvement and optimization as achieved in [2,9,40]. The RS of −51.49 dBm (CB), −51.56 dB (MCB), and −50.25 dBm (GA) is presented for G = 30 dB, l_{fso}= 2500 m, at target BER of 10^{-12} [Proposed work]. Figure 5 shows the BER against the optical SNR (OSNR) (dB) for the WDM-PON/HFFSO link with hybrid OOK/M-ary DPPM- and M-PPM using the CL M = 5, l_{fso} = 4000 m, and D_{RX} = 25 mm for ST. The hybrid OOK/M-ary DPPM-PAPM offers about 4 dB, 6 dB, and 8 dB OSNR improvements over 3-DPPM, 4-DPPM, and 5-DPPM of the ST for the WDM.

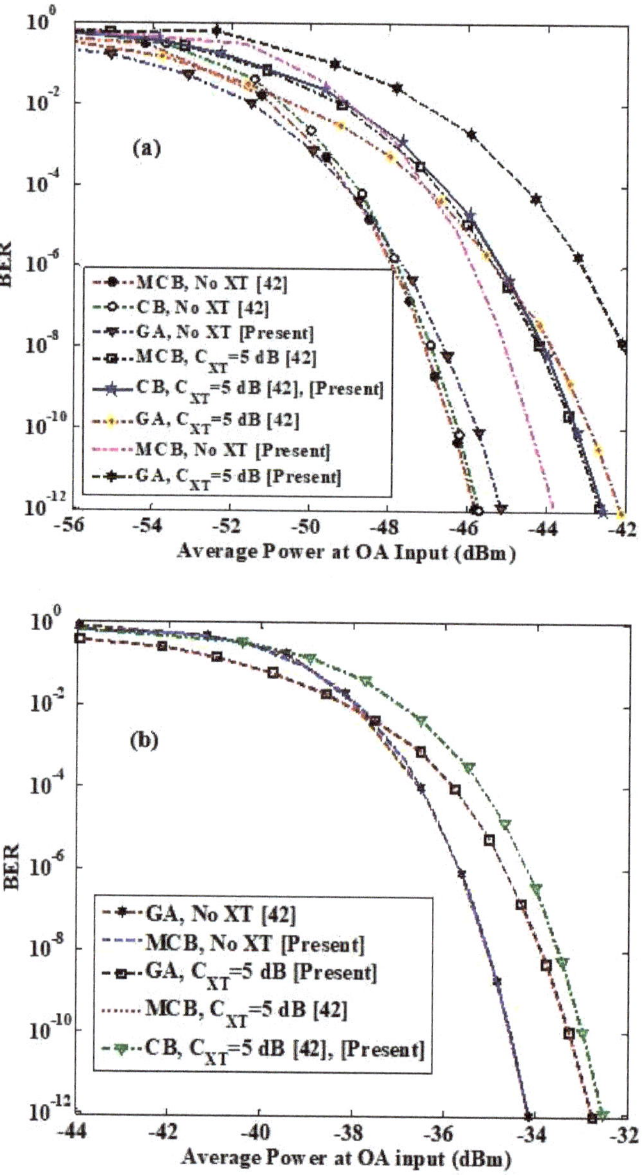

Figure 3. BER against average optical power (AOP) (dBm) using for (**a**) G = 27 dB [42] at 1500 m and G = 30 dB [present work] at 4000 m and (**b**) G = 8 dB ST is demonstrated in [42] at 1500 m and [present work] at 4000 m.

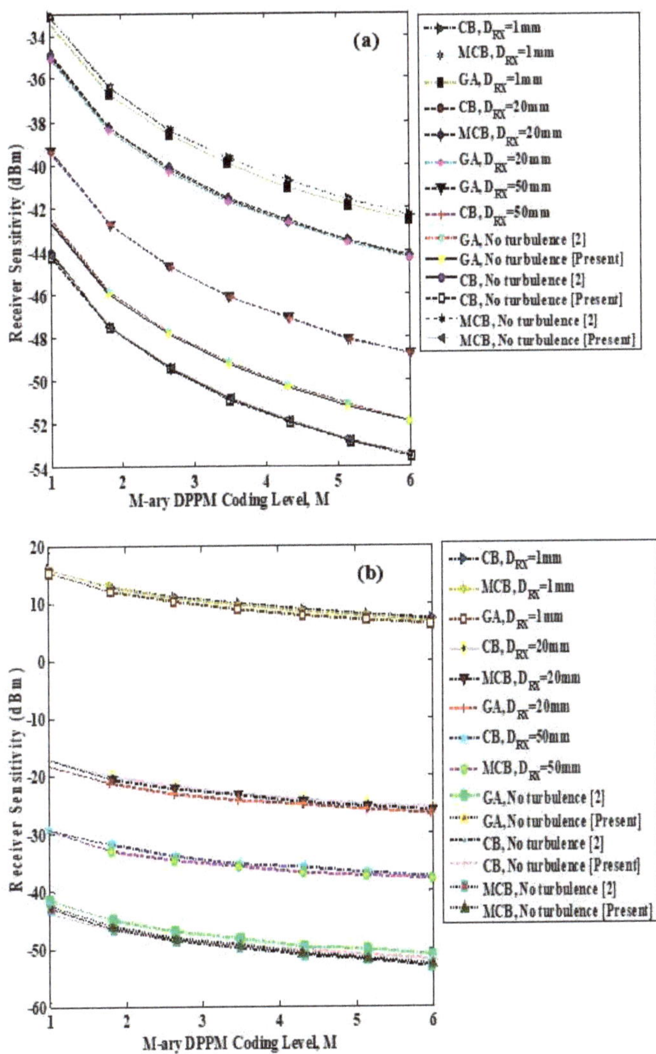

Figure 4. Receiver sensitivity (RS) (dBm) versus M-ary DPPM coding level is presented for the NT with (**a**) WT condition and (**b**) ST is demonstrated in [2], [present].

HFFSO optical fiber network, respectively at a target BER of 10^{-12}. The ST without the hybrid OOK/M-ary modulation format required greater power under the AT effects, ICC, and ASE noise, to achieve the same BER performance than ST with the hybrid OOK/M-ary DPPM scheme as shown in Figure 5. This suggests that the hybrid modulation scheme OOK/M-ary DPPM-PAPM based on WDM-HFFSO communication is favored to ensure greater efficiency and greater performance in our design. The target BER of 10^{-10} is achieved for values of m ≥ 100, while the BER becomes worse specifically at higher ICC and lower CL for value m ≤ 100, as shown in Figure 6. For target BER of 10^{-9}, the resulting DPPM PP analysis for ICC M = 2 is compared to the OOK as shown in Figure 7. The M-DPPM-M-PAPM combination produces less PP than the OOK. The M-DPPM-M-PAPM improvement in the PP is improved as the CL number increases from M = 1 to 2 and the ICC sources [5,8,9,46–56]. The proposed M-ary DPPM-M-PAPM approach increases receiver sensitivity in the WDM-M-HFFSO systems compared with OOK-NRZ and improves reliability and produces a lower PP of 0.2–3.0 dB PP for WT. Without adaptation to reach a BER of 10^{-12}, the modulation techniques decreased OSNR by 4 dB as shown in Figure 8.

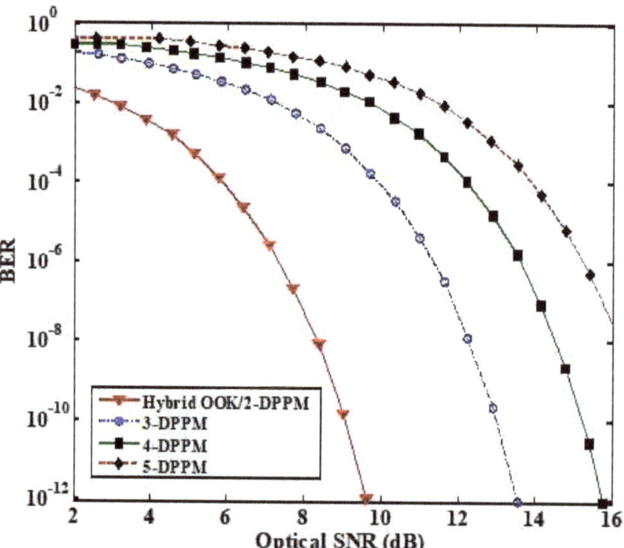

Figure 5. BER vs. OSNR (dB) for the WDM-PON/HFFSO link with hybrid OOK/M-ary DPPM and M-DPPM using D_{RX} = 25 mm, M = 5, l_{fso} = 4000 m for ST.

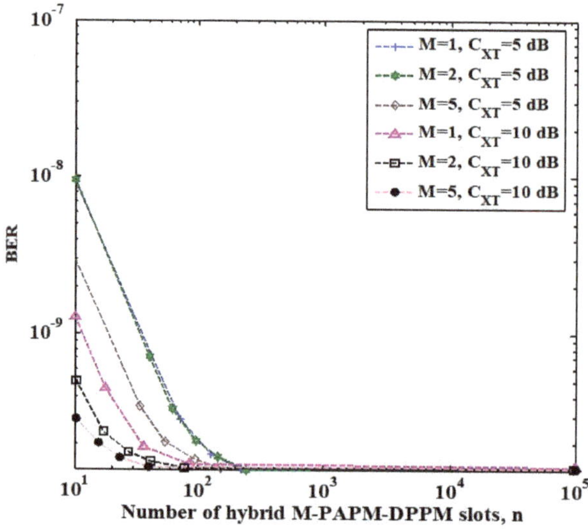

Figure 6. BER vs. the slot numbers (n) for the hybrid M-DPPM-M-PAPM using MCB, G = 30 dB.

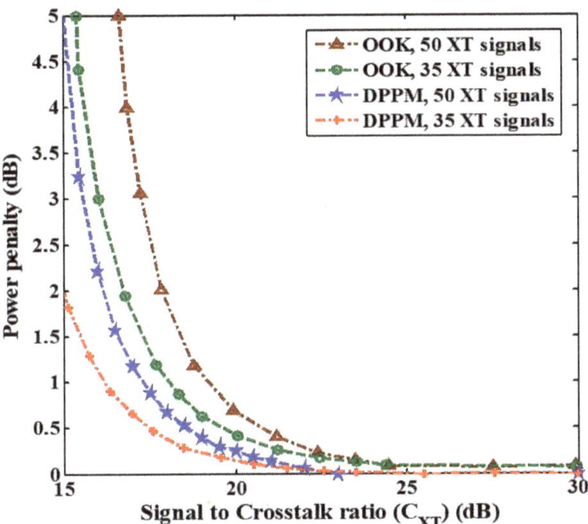

Figure 7. PP vs. C_{XT} for OOK comparison with DPPM-M-PAPM at M = 2 and OOK/M-DPPM-M-PAPM at BER = 10^{-9} for XT (crosstalk) signals 35 and 50.

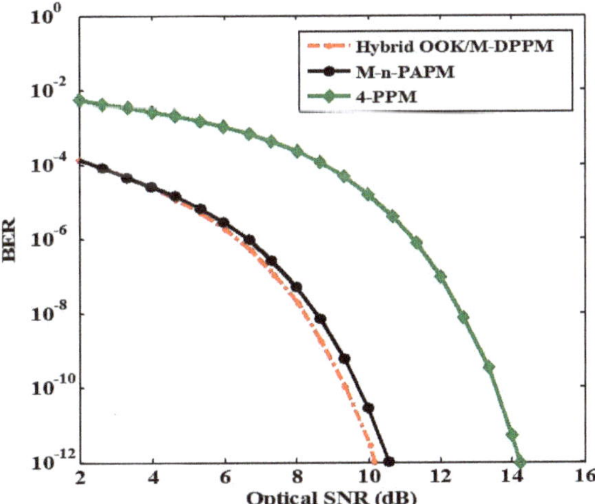

Figure 8. Optical signal-to-noise ratio (OSNR) to BER performance for the hybrid OOK/M-ary DPPM and M-n-PAPM modulation scheme.

Compared to the hybrid OOK/M-ary DPPM, the M-n-PAPM requires 0.3 dB more power. The results demonstrate that the proposed hybrid OOK/M-DPPM and M-n-PAPM models for the WDM-FSO systems can be improved [54–60]. Table 2 summarizes the proposed modulation schemes of MPPM, M-ary-PAPM, and hybrid M-ary PAPM-MPPM with hybrid k-level n-pulse where k^n (with $k \in \{1, 2, \dots \}$) [2,8,9,17,42,50]. Figure 9 shows BER vs. the average RCL power input (dBm), D_{RX} = 25 mm for M-ary DPPM and OOK for NT and (a) WT (b) MT and (c) ST for M-ary DPPM and G = 30 dB, M = 5. At the BER target of 10^{-12}, DPPM provides around 10–11 dB increases in sensitivity compared with the WDM-FSO OOK-NRZ as shown in Figure 9a [2,9,10,17,42–47,56–62]. With a disability in the AT, DPPM's overall sensitivity is reduced to approximately 10 dB (WT), 8 dB (MT), and 8 dB, respectively (ST). The OOK-NRZ based FSO system offers about 7–9 dB for the

DPPM scheme as shown in Figure 9b [2,9]. With AT deficiency, DPPM is reduced to around 7 dB (WT), 8 dB (MT) and 8 dB (ST) [2,9,10,17,42–47], respectively as shown in Figure 9c.

Table 2. The proposed M-ary DPPM-M-PAPM based modulation schemes DWDM-FSO optical fiber communication network system [2,8,9,17,42,50].

Modulation Scheme	Bits per Symbol	Bandwidth Requirement	Spectral Efficiency
MPPM	$\left\lfloor \log_2 \binom{M}{n} \right\rfloor$	$\dfrac{MR_b}{\left\lfloor \log_2 \binom{M}{n} \right\rfloor}$	$\dfrac{\left\lfloor \log_2 \binom{M}{n} \right\rfloor}{M}$
M-ary-PAPM	$\left\lfloor \log_2 k^n \binom{M}{n} \right\rfloor$	$\dfrac{MR_b}{\left\lfloor \log_2 \sum_{i=1}^{n} \binom{M}{n} \right\rfloor}$	$\dfrac{\left\lfloor \log_2 k^n \binom{M}{n} \right\rfloor}{M}$
Hybrid M-ary PAPM-	$\left\lfloor \log_2 k^n \sum_{i=1}^{n} \binom{M}{i} \right\rfloor$	$\dfrac{MR_b}{\left\lfloor \log_2 k^n \sum_{i=1}^{n} \binom{M}{i} \right\rfloor}$	$\dfrac{\left\lfloor \log_2 k^n \sum_{i=1}^{n} \binom{M}{i} \right\rfloor}{M}$

M-ary PAPM-DPPM modulation provides about 10–11 dB increases in OOK-NRZ sensitivity for the WDM-FSO optical fiber network system in the absence of turbulence on the target BER with all AA as seen in Figure 10 [48–58]. When the turbulence and ASE noise is present, the OS is reduced to 10 dB (WT), 8 dB (MT), and 8 dB (ST) [present] in the sensitivity increase of PAPM DPPM over OOK NRZ [56–62]. It is necessary to ensure and continuously maintain the consistency and accuracy of the simulation models and algorithms with reality. This is accomplished mainly by comparing the simulation behavior to the simulation and experimental results achieved in [2,5,8,9,15,42,50]. Also, in our research several (successful) comparisons were done, sets of the previous research and comparing them with the results of models. Tables 1, 3 and 4 show the system parameters used for calculations [2,5,8,9,15,42,50] and [Proposed work]. The simulation results reveal that the BER's WDM/PAPM-DPPM is better than that of OOK-NRZ by approximately 10 dB–11 dB of 10^{-12}. Table 3 provides the performance comparison of the proposed hybrid M-ary-PAPM M-DPPM-based DWDM FSO optical fiber network link with contemporary literature for the references [2,5,8,9,15,42]. The numerical results show a highly efficient M-ary-PAPM M-DPPM-based DWDM FSO has been designed by incorporating a DWDM-PON optical fiber network. The results of the numerical investigation demonstrate that the proposed link performance improves on increasing the increased receiver sensitivity, capacity, and efficiency under the atmospheric turbulence effects. We illustrate a comparative investigation of the proposed M-ary DPPM-M-PAPM-based DWDM-FSO optical fiber communication systems performance with contemporary literature and show that the proposed link performs better for maximum range, efficiency, and channel capacity. We reduce the ASE noise, AT channels, and (ICC for the DWDM-FSO optical fiber communication systems. Table 4 shows the performance comparison of the proposed M-ary DPPM-M-PAPM based-DWDM-FSO optical fiber link with reference [50]. The proposed hybrid M-ary-PAPM M-DPPM provides an improvement of 4–8 dB at a BER of 10^{-12} (OSNR) [proposed work] while in Ref. [50] it is 1 and 2 dB at a BER of 10^{-6}.

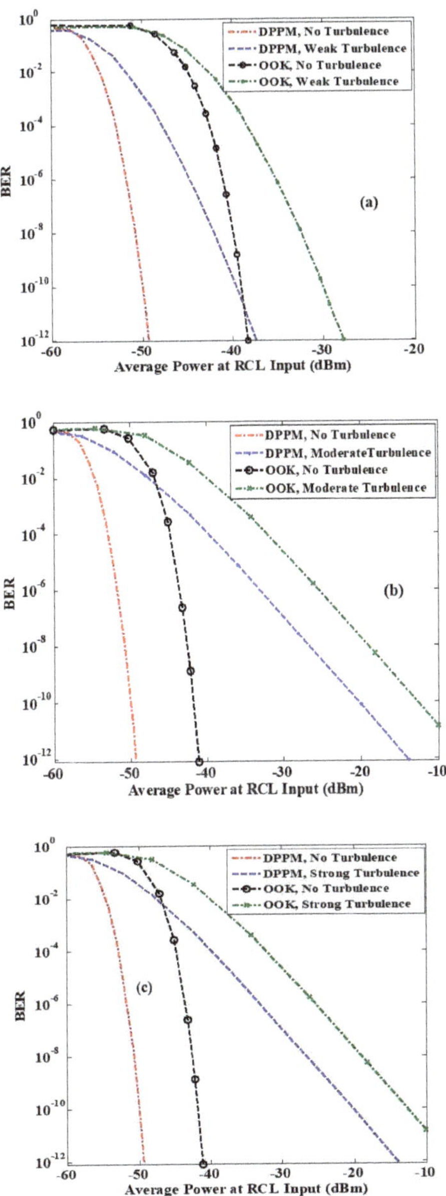

Figure 9. BER vs. AOP at RCL input (dBm) for M-ary DPPM and OOK using D_{RX} = 25 mm, G = 30 dB, M = 5, and l_{fso} = 4000 m for NT with (**a**) WT, (**b**) MT, and (**c**) ST.

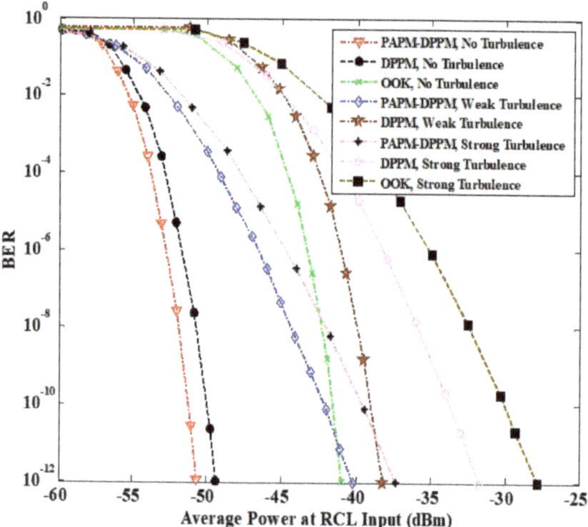

Figure 10. BER vs. AOP at RCL input (dBm) for DPPM and OOK for no turbulence and WT with all apertures averaging (AA).

Table 3. Performance comparison of the proposed M-ary DPPM-M-PAPM based-DWDM-FSO optical fiber link with contemporary literature.

	Ref. [2]	Ref. [5]	Ref. [8]	Ref. [9]	Ref. [15]	Ref. [42]	[Proposed Work]
Technique used	DPPM-FSO link	WDM-FSO hybrid fiber-based DPPM	DPPM- based DWDM-FSO system	M-ary PPM-based WDM -FSO system	WDM-FSO network-based DPPM modulation	DPPM-based WDM-FSO communication system	hybrid M-ary-PAPM M-DPPM-based DWDM FSO optical fiber network
Data rate	2.5 Gbps	2.5 Gbps	2.5 Gbps	2.5 Gbps	2.5 Gbps	2.5 Gbps	2.5 Gbps
Net capacity	2.5 Gbit/s	5 Gbit/s	20 Gbit/s	20 Gbit/s	5 Gbit/s	5 Gbit/s	20 Gbit/s
Maximum link range reported	1500 m	2000 m	4000 m	2500 m	2000 m	2000 m	4000 m
Quantum efficiency	0.75	0.8	1	1	0.8	0.9	1
No. of channels	1	2	8	8	2	2	8
Target BER	10^{-12}	10^{-12}	10^{-12}	10^{-12}	10^{-12}	10^{-12}	10^{-12}

Table 4. Performance comparison of proposed M-ary DPPM-M-PAPM based-DWDM-FSO optical fiber link with reference [50].

	Ref. [50]	[Proposed Work]
Modulation scheme technique	hybrid MPAPM technique deep space optical link	hybrid M-ary-PAPM M-DPPM-based DWDM FSO optical fiber network
OSNR improved	1 dB and 2 dB	4–8 dB
Target BER	10^{-6}	10^{-12}
Optical pulses per frame for M	16	16

6. Conclusions

In this work, we studied and enhanced the HFFSO optical fiber communication network system using the hybrid modulation techniques of OOK/M-ary DPPM-M-PAPM, based on the DWDM-PON network. ICC analyses for DWDM-DPPM systems are given for the GA, CB, and MCB. Also, we consider using hybrid OOK/M-ary DPPM-M-PAPM modulation techniques to increase spectral efficiency and incorporate adaptive optics to mitigate the crosstalk interferences in the DWDM-PON/HFFSO scenarios for improved reliability. The BER performances are then theoretically analyzed. In the presence of atmospheric turbulences, the OOK/M-ary hybrid DPPM modulation scheme is an excellent way to increase DWDM-PON/HFFSO efficiency. Furthermore, the results obtained show that the ICC interferences could be effectively suppressed thanks to the M-ary DPPM-M-PAPM modulation and that the proposed system could achieve superior BER performance. We investigate the impact of the turbulence effect on the proposed system performance based on OOK-M-ary-DPPM modulation as a function of M and other atmospheric parameters. The proposed design of the M-ary DPPM-M-PAPM can improve the power penalty over OOK-NRZ and enhances performance efficiency. The proposed M-ary DPPM-M-PAPM architecture can enhance the receiver sensitivity and reliability over OOK-NRZ.

Author Contributions: M.R.H. made substantial contributions to the design, analysis, characterization, conceptualization, methodology, software, data curation, writing-original draft, formal analysis, writing, visualization, investigation, discussed the results, reviewed, approved the article, validation and provided the revised the article critically for important intellectual content and gave final approval of the version to be submitted. B.B.Y. and M.A.A. participated in the conception and critical revision of the article for important intellectual content, supervision, discussed the results, reviewed and approved the article, project administration, data curation, and writing-review & editing and revised it critically for important intellectual content. All authors have read and agreed to the published version of the manuscript.

Funding: This research received no external funding.

Data Availability Statement: The data that support the findings of this study are available within this article.

Conflicts of Interest: The authors declare that there is no conflict of interest regarding the manuscript.

References

1. Phillips, A.; Cryan, R.; Senior, J. An optically preamplified intersatellite PPM receiver employing maximum likelihood detection. *IEEE Photon-Technol. Lett.* **1996**, *8*, 691–693. [CrossRef]
2. Aladeloba, A.; Phillips, A.; Woolfson, M. Performance evaluation of optically preamplified digital pulse position modulation turbulent free-space optical communication systems. *IET Optoelectron.* **2012**, *6*, 66–74. [CrossRef]
3. Ohtsuki, T. Performance analysis of indoor infrared wireless systems using PPM CDMA. *Electron. Commun. Jpn. Part I Commun.* **2002**, *85*, 1–10. [CrossRef]
4. Leeson, M. Pulse position modulation for spectrum-sliced transmission. *IEEE Photon-Technol. Lett.* **2004**, *16*, 1191–1193. [CrossRef]
5. Mbah, A.M.; Walker, J.G.; Phillips, A.J. Performance evaluation of turbulence-accentuated interchannel crosstalk for hybrid fibre and free-space optical wavelength-division-multiplexing systems using digital pulse-position modulation. *IET Optoelectron.* **2016**, *10*, 11–20. [CrossRef]
6. Garrett, I. Pulse-position modulation for transmission over optical fibers with direct or heterodyne detection. *IEEE Trans. Commun.* **1983**, *31*, 518–527. [CrossRef]
7. Phillips, A.; Cryan, R.; Senior, J. Optically preamplified pulse-position modulation for fibre-optic communication systems. *IEE Proc.-Optoelectron.* **1996**, *143*, 153–159. [CrossRef]
8. Elsayed, E.E.; Yousif, B.B.; Alzalabani, M.M. Performance enhancement of the power penalty in DWDM FSO communication using DPPM and OOK modulation. *Opt. Quantum Electron.* **2018**, *50*, 282. [CrossRef]
9. Elsayed, E.E.; Yousif, B.B. Performance enhancement of M-ary pulse-position modulation for a wavelength division multiplexing free-space optical systems impaired by interchannel crosstalk, pointing error, and ASE noise. *Opt. Commun.* **2020**, *475*, 126219. [CrossRef]
10. Elsayed, E.E.; Yousif, B.B. Performance enhancement of the average spectral efficiency using an aperture averaging and spatial-coherence diversity based on the modified-PPM modulation for MISO FSO links. *Opt. Commun.* **2020**, *463*, 125463. [CrossRef]
11. De Andrade, M.; Kramer, G.; Wosinska, L.; Chen, J.; Sallent, S.; Mukherjee, B. Evaluating strategies for evolution of passive optical networks. *IEEE Commun. Mag.* **2011**, *49*, 176–184. [CrossRef]
12. Kramer, G.; Pesavento, G. Ethernet passive optical network (EPON): Building a next-generation optical access network. *IEEE Commun. Mag.* **2002**, *40*, 66–73. [CrossRef]
13. Kim, B.; Kim, B.-W. WDM-PON development and deployment as a present optical access solution. In Proceedings of the Optical Fiber Communication Conference, San Diego, CA, USA, 22–26 March 2009.
14. Wang, K.; Nirmalathas, A.; Lim, C.; Skafidas, E. 4 × 12.5 Gb/s WDM optical wireless communication system for indoor applications. *J. Light. Technol.* **2011**, *29*, 1988–1996. [CrossRef]
15. Aladeloba, A.O.; Woolfson, M.S.; Phillips, A.J. WDM FSO network with turbulence-accentuated interchannel crosstalk. *J. Opt. Commun. Netw.* **2013**, *5*, 641–651. [CrossRef]

16. Chang, G.-K.; Chowdhury, A.; Jia, Z.; Chien, H.-C.; Huang, M.-F.; Yu, J.; Ellinas, G. Key technologies of WDM-PON for future converged optical broadband access networks. *J. Opt. Commun. Netw.* **2009**, *1*, C35–C50. [CrossRef]
17. Available online: http://eprints.nottingham.ac.uk/13304/1/AladelobaAbisayoThesis.pdf (accessed on 10 October 2021).
18. Ciaramella, E.; Arimoto, Y.; Contestabile, G.; Presi, M.; D'Errico, A.; Guarino, V.; Matsumoto, M. 1.28 terabit/s (32 × 40 Gbit/s) WDM transmission system for free space optical communications. *IEEE J. Sel. Areas Commun.* **2009**, *27*, 1639–1645. [CrossRef]
19. Forbes, M.; Gourlay, J.; Desmulliez, M. Optically interconnected electronic chips: A tutorial and review of the technology. *Electron. Commun. Eng. J.* **2001**, *13*, 221–232. [CrossRef]
20. Ansari, N.; Zhang, J. *Media Access Control and Resource Allocation: For Next Generation Passive Optical Networks*; Springer Science & Business Media: Berlin/Heidelberg, Germany, 2013.
21. Zuo, T.; Phillips, A. Performance of burst-mode receivers for optical digital pulse position modulation in passive optical network application. *IET Optoelectron.* **2009**, *3*, 123–130. [CrossRef]
22. Yousif, B.B.; Elsayed, E.E. Performance enhancement of an orbital-angular-momentum-multiplexed free-space optical link under atmospheric turbulence effects using spatial-mode multiplexing and hybrid diversity based on adaptive MIMO equalization. *IEEE Access* **2019**, *7*, 84401–84412. [CrossRef]
23. Yamamoto, Y. Noise and error rate performance of semiconductor laser amplifiers in PCM-IM optical transmission systems. *IEEE J. Quantum Electron.* **1980**, *16*, 1073–1081. [CrossRef]
24. Ribeiro, L.; da Rocha, J.; Pinto, O.; da Rocha, J.F.; Pinto, J. Performance evaluation of EDFA preamplified receivers taking into account intersymbol interference. *J. Light. Technol.* **1995**, *13*, 225–232. [CrossRef]
25. O'Reilly, J.; da Rocha, J.F. Improved error probability evaluation methods for direct detection optical communication systems. *IEEE Trans. Inf. Theory* **1987**, *33*, 839–848. [CrossRef]
26. Al-Orainy, A.; O'Reilly, J. Error probability bounds and approximations for the influence of crosstalk on wavelength division multiplexed systems. *IEE Proc. J Optoelectron.* **1990**, *137*, 379–384. [CrossRef]
27. Personick, S.D. Applications for quantum amplifiers in simple digital optical communication systems. *Bell Syst. Tech. J.* **1973**, *52*, 117–133. [CrossRef]
28. Ma, R.; Zuo, T.; Phillips, A.; Sujecki, S. Improved performance evaluation for DC-coupled burst mode reception in the presence of amplified spontaneous emission noise and interchannel crosstalk. *IET Optoelectron.* **2010**, *4*, 121–132. [CrossRef]
29. Elsayed, E.E.; Yousif, B.B. Performance enhancement of hybrid diversity for M-ary modified pulse-position modulation and spatial modulation of MIMO-FSO systems under the atmospheric turbulence effects with geometric spreading. *Opt. Quantum Electron.* **2020**, *52*, 1–18. [CrossRef]
30. Andrews, L.C.; Phillips, R.L.; Young, C.Y. *Laser Beam Scintillation with Applications*; SPIE: Bellingham, WA, USA, 2001. [CrossRef]
31. Trinh, P.V.; Dang, N.T.; Thang, T.C.; Pham, A. Performance of all-optical amplify-and-forward WDM/FSO relaying systems over atmospheric dispersive turbulence channels. *IEICE Trans. Commun.* **2016**, *99*, 1255–1264. [CrossRef]
32. Majumdar, A.K. Free-space laser communication performance in the atmospheric channel. *J. Opt. Fiber Commun. Rep.* **2005**, *2*, 345–396. [CrossRef]
33. Khalighi, M.; Schwartz, N.; Aitamer, N.; Bourennane, S. Fading reduction by aperture averaging and spatial diversity in optical wireless systems. *J. Opt. Commun. Netw.* **2009**, *1*, 580–593. [CrossRef]
34. Al-Habash, M.A. Mathematical model for the irradiance probability density function of a laser beam propagating through turbulent media. *Opt. Eng.* **2001**, *40*, 1554. [CrossRef]
35. Popoola, W.O.; Ghassemlooy, Z. BPSK subcarrier intensity modulated free-space optical communications in atmospheric turbulence. *J. Light. Technol.* **2009**, *27*, 967–973. [CrossRef]
36. Rajiv Ramaswami, K.N.S. Optical Networks a Practical Perspective. 2010. Available online: http://www.cesarkallas.net/arquivos/faculdade-pos/TP319-redes-opticas/Optical-Networks-3nd.pdf (accessed on 10 October 2021).
37. Maru, K.; Mizumoto, T.; Uetsuka, H. Demonstration of flat-passband multi/demultiplexer using multi-input arrayed waveguide grating combined with cascaded mach–zehnder interferometers. *J. Light. Technol.* **2007**, *25*, 2187–2197. [CrossRef]
38. Hirano, A.; Miyamoto, Y.; Kuwahara, S. Performances of CSRZ-DPSK and RZ-DPSK in 43-Gbit/s/ch DWDM G.652 single-mode-fiber transmission. *J. Lightwave Technol.* **2003**, *86*, 454–456. [CrossRef]
39. Yu, C.X.; Neilson, D.T. Diffraction-grating-based (de)multiplexer using image plane transformations. *IEEE J. Sel. Top. Quantum Electron.* **2002**, *8*, 1194–1201. [CrossRef]
40. Henry, P.S. Error-rate performance of optical amplifiers. In Proceedings of the Optical Fiber Communication Conference, Houston, TX, USA, 6 February 1989.
41. Elsayed, E.E.; Yousif, B.B. Performance evaluation and enhancement of the modified OOK based IM/DD techniques for hybrid fiber/FSO communication over WDM-PON systems. *Opt. Quantum Electron.* **2020**, *52*, 1–27. [CrossRef]
42. Mbah, A.M.; Walker, J.G.; Phillips, A.J. Performance evaluation of digital pulse position modulation for wavelength division multiplexing FSO systems impaired by interchannel crosstalk. *IET Optoelectron.* **2014**, *8*, 245–255. [CrossRef]
43. Yousif, B.B.; Elsayed, E.E.; Alzalabani, M.M. Atmospheric turbulence mitigation using spatial mode multiplexing and modified pulse position modulation in hybrid RF/FSO orbital-angular-momentum multiplexed based on MIMO wireless communications system. *Opt. Commun.* **2019**, *436*, 197–208. [CrossRef]
44. Mbah, A.M.; Walker, J.G.; Phillips, A.J. Outage probability of WDM free-space optical systems affected by turbulence-accentuated interchannel crosstalk. *IET Optoelectron.* **2017**, *11*, 91–97. [CrossRef]

45. Aladeloba, A.; Phillips, A.; Woolfson, M. Improved bit error rate evaluation for optically pre-amplified free-space optical communication systems in turbulent atmosphere. *IET Optoelectron.* **2012**, *6*, 26–33. [CrossRef]
46. Aladeloba, A.O.; Phillips, A.J.; Woolfson, M.S. DPPM FSO communication systems impaired by turbulence, pointing error and ASE noise. In Proceedings of the 14th International Conference on Transparent Optical Networks (ICTON), Coventry, UK, 2–5 July 2012; pp. 1–4. [CrossRef]
47. Mukherjee, B. WDM optical communication networks: Progress and challenges. *IEEE J. Sel. Areas Commun.* **2000**, *18*, 1810–1824. [CrossRef]
48. Mallick, K.; Mandal, P.; Mandal, G.C.; Mukherjee, R.; Das, B.; Patra, A.S. Hybrid MMW-over fiber/OFDM-FSO transmission system based on doublet lens scheme and POLMUX technique. *Opt. Fiber Technol.* **2019**, *52*, 101942. [CrossRef]
49. Mallick, K.; Mandal, P.; Mukherjee, R.; Mandal, G.C.; Das, B.; Patra, A.S. Generation of 40 GHz/80 GHz OFDM based MMW source and the OFDM-FSO transport system based on special fine tracking technology. *Opt. Fiber Technol.* **2020**, *54*, 102130. [CrossRef]
50. Idris, S.; Selmy, H.; Lopes, W.T.A. Performance analysis of hybrid MPAPM technique for deep-space optical communications. *IET Commun.* **2021**, *15*, 1700–1709. [CrossRef]
51. Magidi, S.; Jabeena, A. Analysis of multi-pulse position modulation free space optical communication system employing wavelength and time diversity over Malaga turbulence channel. *Sci. Afr.* **2021**, *12*, e00777. [CrossRef]
52. Alipour, A.; Farmani, A.; Mir, A. Analysis of optical power budget in DWDM-FSO link under outdoor atmospheric channel model. *Opt. Quantum Electron.* **2021**, *53*, 1–15. [CrossRef]
53. Ran, H.; Zhang, J.; Pan, G.; Xie, Y. Outage probability of wireless-powered multi-relaying MIMO FSO-RF systems. *Opt. Commun.* **2021**, *498*, 127260. [CrossRef]
54. Zhang, T.; Wang, P.; Liu, T.; Jia, C.; Pang, W.-N.; Wang, W. Performance analysis of multi-hop parallel FSO system over double generalized gamma distribution considering two transmission beams. *Optoelectron. Lett.* **2021**, *17*, 215–220. [CrossRef]
55. Willner, A.E.; Zhao, Z.; Liu, C.; Zhang, R.; Song, H.; Pang, K.; Manukyan, K.; Song, H.; Su, X.; Xie, G.; et al. Perspectives on advances in high-capacity, free-space communications using multiplexing of orbital-angular-momentum beams. *APL Photonics* **2021**, *6*, 030901. [CrossRef]
56. Sharma, K.; Grewal, S.K. Performance assessment of hybrid PPM–BPSK–SIM based FSO communication system using time and wavelength diversity under variant atmospheric turbulence. *Opt. Quantum Electron.* **2020**, *52*, 1–25. [CrossRef]
57. Singh, M.; Malhotra, J. Performance comparison of M-QAM and DQPSK modulation schemes in a 2 × 20 Gbit/s–40 GHz hybrid MDM–OFDM-based radio over FSO transmission system. *Photon.-Netw. Commun.* **2019**, *38*, 378–389. [CrossRef]
58. Malik, S.; Sahu, P.K. Performance analysis of free space optical communication system using different modulation schemes over weak to strong atmospheric turbulence channels. *Lect. Notes Electr. Eng.* **2019**, *546*, 387–399. [CrossRef]
59. Srivastava, V.; Mandloi, A.; Soni, G.G. Outage probability and average BER estimation of FSO system employing wavelength diversity. *Opt. Quantum Electron.* **2019**, *51*, 229. [CrossRef]
60. Mukherjee, R.; Mallick, K.; Kuiri, B.; Santra, S.; Dutta, B.; Mandal, P.; Patra, A.S. PAM-4 based long-range free-space-optics communication system with self injection locked QD-LD and RS codec. *Opt. Commun.* **2020**, *476*, 126304. [CrossRef]
61. Mukherjee, R.; Mallick, K.; Mandal, P.; Dutta, B.; Kuiri, B.; Patra, A.S. Bidirectional hybrid OFDM based free-space/wireless-over-fiber transport system. *Opt. Quantum Electron.* **2020**, *52*, 1–12. [CrossRef]
62. Saxena, P.; Mathur, A.; Bhatnagar, M.R. BER performance of an optically pre-amplified FSO system under turbulence and pointing errors with ASE noise. *J. Opt. Commun. Netw.* **2017**, *9*, 498–510. [CrossRef]

 *photonics*

Article

Patent Technology Network Analysis of Machine-Learning Technologies and Applications in Optical Communications

Shu-Hao Chang [ID]

Science and Technology Policy Research and Information Center, National Applied Research Laboratories, Taipei 10663, Taiwan; shchang@narlabs.org.tw; Tel.: +866-2-27377779

Received: 12 October 2020; Accepted: 14 December 2020; Published: 15 December 2020

Abstract: As the Internet of Things (IoT) develops, applying machine learning on optical communications has become a prospective field of research. Scholars have mostly concentrated on algorithmic techniques or specific applications but have been unable to address the distribution of machine-learning technologies and the development of its applications in optical communications from a macro perspective. Therefore, in this paper, machine-learning patents in optical communications are taken as the analytical basis for constructing a patent technology network. The study results revealed that key technologies were primarily in data input and output devices, data-processing methods, wireless communication networks, and the transmission of digital information in optical communications. Such technologies were also applied to perform measurement for diagnostic purposes and medical diagnoses. The technology network model proposed in this paper explores the technological development trends of machine learning in optical communications and serves as a reference for allocating research and development resources.

Keywords: machine learning; optical communications; patent analysis; network analysis; technical analysis

1. Introduction

Optical communications have greatly advanced in signal serial communication speed, agile channel spacing, modulation formats, and coding schemes. However, relevant technologies have yet to fully meet the complexity and performance requirements of future optical communication system networks. The distinguishing features of machine learning are its autonomous learning and evolution ability. With new information, a system can modulate its structure and parameters to build a new mapping network and enable new skills [1]. At present, machine-learning technologies play a significant role in network planning, failure prediction, and optical performance monitoring in optical communication systems [2–4]. In the future, intelligent optical communication system networks will be automated and adaptive and become capable of predicting traffic demands to maximize performance. To achieve this goal, the integration of machine-learning mathematics, programming, and algorithms is necessary in optical communications, and these are the key directions of future optical communication development.

Integrating machine learning and optical communication technologies—in essence, combining computer science with communication—is a forward-looking research field. Machine-learning technologies have been highly effective in classification tasks, particularly when signals are non-linear and distorted [5]. Machine learning predicts and eliminates defects in a system by learning its properties. Current signal analysis systems are ineffective in classifying signals. However, machine learning can be applied to such systems to identify patterns in collected data and boost the systems' signal analysis performance. In truth, more scholars have begun to focus on developing this technology [2,4,6].

Studies have mainly focused on algorithmic techniques [4,7] or specific applications [2,8]; however, these studies have not addressed the technological distribution and application development of machine learning in optical communication from a macro perspective. In particular, the rapid development of machine learning has gradually rendered relevant technologies a powerful tool for enterprises in smart manufacturing to increase output value. The development of the Internet of Things (IoT) and its high-speed, low-latency, and large-data transmission characteristics will drive optical communication needs. Integrating the machine-learning and optical communication technologies will unlock unlimited business opportunities in the future; thus, defining key technologies and their applications are critical. In the current phase of machine learning, where personnel and funds should be directed, is the most crucial consideration in optical communications for government and industry actors. Machine learning was applied for interface arrangements, image capture, encoding, decoding, and code conversion. Governments may consider investing more research funds and resources toward academic research for developing technologies in this field. This study addressed this topic through employing technology network analysis to analyze and identify key technologies.

This paper focuses on technology networks for machine learning in optical communications and constructs a technology network model through patent analysis. Patents are the most direct method of measuring the industrialization of technology, and previous measurements of technology transfers [9,10] and studies on technology development trajectories [11,12] and industrial technology development maps [13,14] have used patents. Therefore, patent information is one of the most direct metrics for technological development. Patent information forms the basis of this paper's observation of technological development trends and key technological fields. In terms of the screening of machine learning and optical communications patents, a prior art search for patents in the past mainly involved the use of keywords [15,16], similarities in International Patent Classification (IPC) codes [17,18], or the establishment of a topic-dependent citation graph [19]. The patents were retrieved using Derwent smart search (SSTO), which involves hundreds of experts who read and translate the open information of official patents in various databases before rewriting the key abstracts, eliminating content errors, and normalizing the patentees. The revised and normalized data are logged into a database after manual browsing and sorting.

This study differs from previous explorations of algorithmic techniques and applications by mainly exploring key technologies and applicable developments for machine learning in optical communications. It focuses on building a technology network model and identifying technological development trends. The study results can serve as a reference for government and industry actors.

2. Literature Review

2.1. Current Development of Machine Learning in Optical Communications

The two vital forces driving the advancement of the overall communications industry are fifth-generation mobile communication system (5G) and artificial intelligence (AI) technologies. To increase communication transmission speeds, the application of optical communications technology has gradually increased in the 5G field. A study by MarketsandMarkets stated, "The global optical communication and networking equipment market size was valued at USD 18.9 billion in 2020 and is projected to reach USD 27.8 billion by 2025; it is growing at a compound annual growth rate (CAGR) of 8.0% from 2020 to 2025" [20]. AI fulfills critical roles in optical communication industries, including optical network planning and operations in transport and access networks [21]. Machine learning is the basis of AI, and studies have argued that machine-learning techniques are used for many optical network applications. The applications of machine learning are classified by their use in cases, which are categorized into optical network control and resource management and optical network monitoring and survivability [22]. Therefore, methods of improving the performance of optical communications through mathematical methods and algorithms related to machine learning may become the trend of future research.

To satisfy the growing traffic demands of mobile communications and transmitting diverse and high-quality data through methods such as IoT, the goal of developing optical communications is to integrate machine learning and create deeply intelligent control management systems. Topics on involving machine learning in optical communications are diverse and include optical performance monitoring, fiber non-linearity compensation, cognitive network failure prediction, dynamic planning, the cross-layer optimization of software-defined networks, the quality of transmission estimation, and the physical layer design of optical communication systems [23]. These techniques involve optics, mathematics, computer science, communications, and semiconductors and belong to cross-disciplinary technical fields. Developments in IoT technologies, mobile communications, and optical communications have led multiple governments to focus on machine-learning applications in optical communications. Optical communications will gradually become widespread among most users. The sending of large volumes of sound, video, and image data is reliant on AI technologies such as machine learning for computing and transmission, and the range of applicable fields is wide. Therefore, this study analyzed the technological development and applicable fields of machine learning in optical communications, and patents were examined to identify key technologies. The exploration into these key technologies was conducted through network analysis, which is further explained in the following section.

2.2. Technology Network Analysis Model

In recent years, studies have explored the development and trends of technological innovation through network analysis methods [24,25] or sought to understand the technical partnership between institutions or inventors through network analysis [26,27] to explore the flow of knowledge [28,29]. Network analysis can be used to accurately illustrate the transmission paths and evolution of technology and knowledge. Analysis of patent data in particular can provide objective and feasible data, such as the year the patent was approved, the quantity of the patent, and the type of technology used [13]. Therefore, this study used patent data to analyze the development of specific technologies, and based on the features of network analysis, the relationship between technical fields were analyzed using co-classification. Because each patent may be involved in multiple technical fields, co-classification can be used to define the relationship between technical fields [30,31] and pinpoint key technologies on the basis of technology networks. The classification framework of technical fields was based on the current the IPC system, and this study used the technology network analysis model to explore key technical fields in machine learning in optical communications.

3. Research Design

3.1. Retrieval Strategy and Data Source

In this study, the patent analysis was based on data from the United States Patent and Trademark Office (USPTO), a historic organization whose development and data can be traced back to 1975. Because the United States is the largest commercial trade market in the world, most inventors who apply for patents in other countries also submit patent applications in the United States. Therefore, researchers generally use the USPTO database to examine global innovation activities [13,14]. Patents explored in this study were limited to US patents that were announced between January 2015 and December 2019. In this study, the patents were retrieved using SSTO, and the search criteria were (SSTO/Machine Learning) and (SSTO/Optical Communications), which returned 824 patents in total.

3.2. Network Centrality Analysis

In this study, key technologies in patent technology networks were identified through technology network centrality. The methods of measuring network centrality are explained as follows.

3.2.1. Degree Centrality

Degree centrality is the number of nodes that are adjacent a specific node and can be used to evaluate the core of a patent technology network. High degree centrality represents a greater number of connected nodes in a network, and degree centrality in specific networks of links represents critical transitions that will become a hot spot in the network [32].

$$C_d(i) = \sum_j m_{ji} \tag{1}$$

If nodes i and j are connected, then $m_{ji} = 1$.

3.2.2. Eigenvector Centrality

Eigenvector centrality measures the influence of a node in a network. In addition to whether a node is connected with other nodes, relevant analysis focuses on whether the nodes connected to this nodes are linked with other nodes. The centrality of a node is determined by the centrality of its adjacent nodes. If a node is connected to nodes with high centrality, the node in question has higher degree centrality. This indicates that degree centrality differs between adjacent nodes. Analysis of eigenvector centrality can determine the relative importance of a node and constitutes a crucial research field in technology networks.

$$C_e(i) = \lambda^{-1} \sum_{j=1}^{n} a_{ij} C_e(j) \tag{2}$$

where $C_e(i)$ and $C_e(j)$ are the eigenvector centrality of nodes i and j, respectively; a_{ij} represents the node entering the adjacency matrix A; and λ is a constant and the largest eigenvalue in the adjacency matrix A.

In this equation, eigenvector centrality views the centrality of a single node as the linear combination of the centrality of all other nodes to derive a linear function [33].

3.2.3. Structural Hole

Structural holes can be used to assess the ability of a node as a mediator in the overall network. This concept describes the characteristics of a node that occupies the main communication and information channels in the network, which is associated with the hole effect [34]. This is the degree to which the connections between clusters depend on the node in question. The gaps between clusters provide opportunities to build a network of bridges. If any technology becomes a bridge to connect two non-overlapping technology clusters, that bridging technology gains a spatial advantage in the overall technology network. Burt [34] argued that structural hole effects can be measured using a network constraint index, with a value between 0 and 1. A high network constraint index indicates low autonomy and a low structural hole effect in the entity.

$$C_{ij} = \left(P_{ij} + \sum_k P_{jk} P_{kj} \right)^2 \tag{3}$$

where C_{ij} is the score of the constraint on node i by node j; P_{ij} is the proportion of connections with node j among the connections of node i; P_{jk} is the relational ratio of node j with the other connections of all nodes; and P_{kj} is the ratio of all other nodes with node j connections.

This equation sums all node j totals, and this sum is the total constraint on node i in the network [34]; ergo, $C_i = \sum_j C_{ij}$.

4. Results

4.1. Patent Retrieval Results

Before performing technology network analysis, the patent retrieval results were analyzed to gain a preliminary overview of technological development. Machine-learning technologies in optical communications involved 407 four-digit IPCs, which indicated a wide scope of involvement. Table 1 presents the quantity of the top 10 four-digit IPCs.

Table 1. Top 10 International Patent Classifications (IPCs) to the fourth digit.

Ranking	IPC Classification	Frequency	Percentage
1	G06F17	170	5.17%
2	G06F3	142	4.31%
3	G06K9	108	3.28%
4	H04L29	102	3.10%
5	A61B5	89	2.70%
6	G06F9	82	2.49%
7	H04L12	75	2.28%
8	G10L15	71	2.16%
9	H04B10	69	2.10%
10	G06T7	58	1.76%

In Table 1, the frequency denotes the number of patents that have appeared in the IPC classification; for example, 170 patents under G06F17 belonged to the IPC classification. The percentage represents the proportion accounted for by an IPC classification out of the total number of IPC classifications; for example, the 824 patents contained 3291 IPC classifications (one patent might have more than two IPC classifications), and G06F17 appeared 170 times, accounting for 5.17%. The results indicated that that the technologies were mostly concentrated under the classifications G06F17, G06F3, G06K9, H04L29, and A61B5 (Table 1); Appendix A displays the definition of each IPC code. According to the IPC definitions, G06F17 refers to data-processing methods, and G06F3 is the classification for data input and output devices (e.g., interface arrangements). G06K9 covers methods and devices for recognizing patterns, and H04L29 is communication control. A61B5 is the classification for measuring diagnostic purposes, among which G06F17 was more related to AI [35] because it appeared the most frequently.

The analysis results of the top ten patentees revealed that Apple Inc., which focuses on AI and communication technology development, holds the greatest number of patents, followed by SAS Institute Inc., Intel Corporation, and International Business Machines Corporation, which are global leaders in intelligent software and services (Table 2). The next patentee, Volcano Corporation, develops biological diagnostic systems with ultrahigh resolution and holds numerous patents to technologies that can be applied for medical diagnostics.

Table 2. Top 10 patentees.

Ranking	Patentee	Patents	Percentage	No. of Inventors
1	Apple Inc.	67	8.15%	142
2	SAS Institute Inc.	51	6.20%	92
3	Intel Corporation	33	4.01%	65
4	International Business Machines Corporation	24	2.92%	67
5	Volcano Corporation	29	3.53%	29
6	Elwha LLC	28	3.41%	33
7	Microsoft Technology Licensing, LLC	17	2.07%	51
8	Google LLC	16	1.95%	42
9	Fitbit, Inc.	14	1.70%	10
10	HPS Investment partners, LLC, as Collateral Agent	13	1.58%	18

Furthermore, G06N in three-digit IPCs describes computer systems based on specific computational models, and G06N will be subdivided into four-digit IPCs, which is further explained different computational models. Therefore, the patents collected in this study reveal computational models (i.e., G06N). In this study, four-digit IPCs related to G06N appeared 131 times, and the distribution of the four-digit IPCs, including computer systems based on biological models (G06N3), computer systems using knowledge-based models (G06N5), subject matter not provided for in other groups of this subclass (G06N99), as shown in Table 3.

Table 3. Computational models classification results.

Methods	Frequency	Percentage
Computer systems based on biological models	36	27.48%
Computer systems using knowledge-based models	28	21.37%
Subject matter not provided for in other groups of this subclass	28	21.37%
Others	39	29.77%

4.2. Technology Network Analysis

The results of previous studies have suggested that technology co-classification analysis can be used to analyze the relationship between fields of technology [30,31]. Because patents may be subject to multiple patent classification codes, co-classification information can be used to define the relationship between technical fields, as shown in Figure 1.

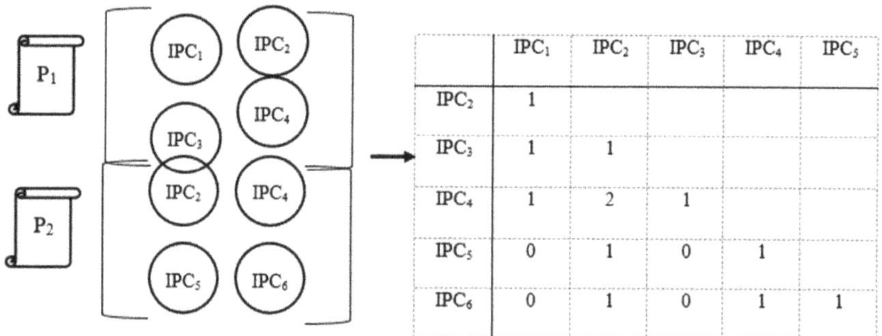

	IPC₁	IPC₂	IPC₃	IPC₄	IPC₅
IPC₂	1				
IPC₃	1	1			
IPC₄	1	2	1		
IPC₅	0	1	0	1	
IPC₆	0	1	0	1	1

Figure 1. Technical field co-classification analysis schematic. Note: P_1 and P_2 represent two patents; IPC_1, IPC_2, $IPC_3 \ldots IPC_6$ represent six technical fields.

The numbers in the matrix in Figure 1 represent the frequencies of different IPCs appearing in the same patent. A greater number represents a stronger technical connection between IPCs. For example, the IPCs of P_1 belong to IPC_1, IPC_2, IPC_3, and IPC_4. Because IPC_1, IPC_2, IPC_3, and IPC_4 simultaneously appear in P_1, a technical connection exists among IPC_1, IPC_2, IPC_3, and IPC_4. Moreover, regarding the relationship between IPC_1 and other technical fields, because IPC_1, IPC_2, IPC_3, and IPC_4 only appear simultaneously in P_1, the technical connections between IPC_1 and IPC_2, IPC_3, and IPC_4 were consistently "1", as shown in the first column of the matrix. Because IPC_1 does not concurrently appear with IPC_5 and IPC_6 in patents, the technical connection of IPC_1 to IPC_5 and IPC_6 is "0". This approach was adopted to gradually develop co-classification matrices of all technical fields. Therefore, the present study can facilitate plotting of the technology network map through the matrix.

Table 4 presents all parameters used in the technology network analysis. Figure 2 presents the network model of key technologies, and the key IPCs are listed in Table 5. The centrality performance index of the top 10 IPC codes in the frequency analysis (Table 1) has been added to Appendix B.

Table 4. Parameters related to technology network analysis.

Items	Technology Network Analysis
Nodes	407
Links	206,194
Average path length	2.688
Network density	0.159
The average degree	16.015
Compactness	0.377

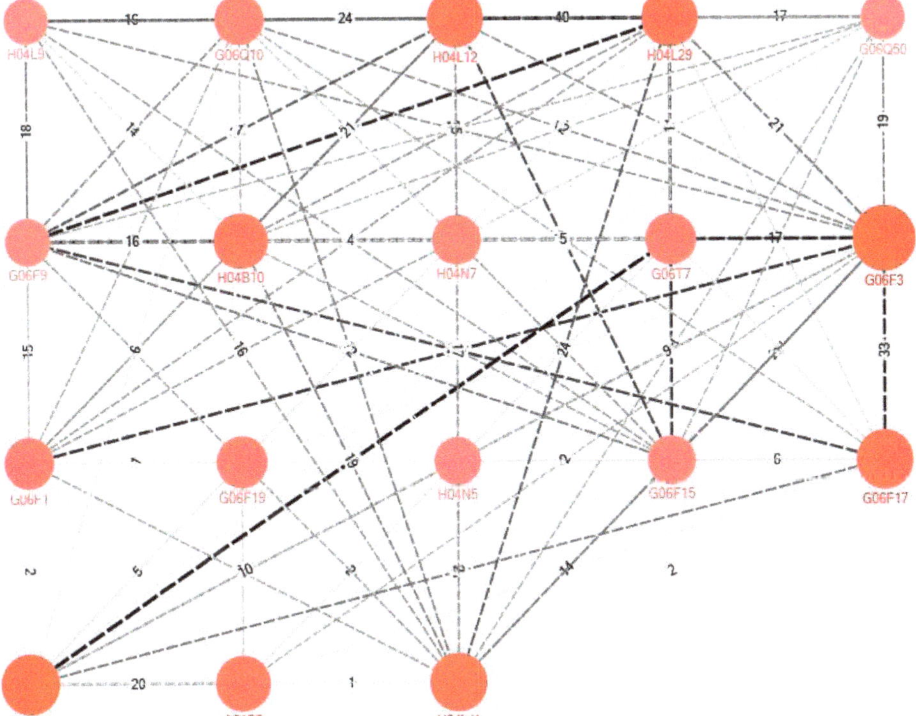

Figure 2. IPC network for machine learning in optical communications. Note: node size represents the number of connections with a node, and the arc thickness represents the strength of a connection. This diagram only displays nodes with 60 or more connections.

Table 5. Top 5 IPC codes for machine learning in optical communications.

IPC	Degree Centrality	IPC	Eigenvector Centrality	IPC	Structural Hole
G06F3	126	G06F3	0.206	G06F3	0.063
G06K9	109	H04W4	0.183	A61B5	0.076
G06F17	102	H04L29	0.182	G06F19	0.077
H04W4	100	H04L12	0.176	H04W4	0.082
H04L12	97	H04B10	0.169	G06F17	0.085

The overall network, as shown in Table 4, was composed of 407 nodes and 206,194 links; a total of 407 IPCs were related to machine learning technologies in optical communications. The network density and compactness were 0.159 and 0.377, respectively, indicating that the network was sparsely distributed and that the interaction frequency between nodes was low. The average path length was 2.688, meaning that connecting a node to other nodes required nearly 3 steps, on average. The technology nodes depicted in Figure 1 are crucial technology nodes linking more than 60 different technology nodes; that is, they represent the more critical technology fields in patents for machine-learning technologies in optical communications. Regarding degree centrality, eigenvector centrality, and structural hole, the IPCs G06F3, G06F17, H04W4, and H04L12 had more than two indices in the top five technology fields (Table 5). In the overall IPC codes, the mean eigenvector centrality was 0.028, and a larger value indicated greater relative importance of a node. The structural hole effect was measured using the network constraint index. The mean of said index was 0.519, and a higher value indicated a lower structural hole effect. This suggested that key machine-learning technologies in optical communications are mainly concentrated in data input and output devices (e.g., interface arrangements; G06F3), data-processing methods (G06F17), wireless communication networks (H04W4), and the transmission of digital information (H04L12). The IPCs A61B5 and G06F19 only appeared in the structural hole column, which signified that these two technology fields belonged to different technology clusters and played a cross-disciplinary role. Thus, key technologies in the cross-disciplinary uses of machine learning in optical communications were for measurements for diagnostic purposes (A61B5) and information and communication technology (ICT) specially adapted for specific application fields (G06F19).

Additional insights available through network centrality metrics were as follows. Despite the slightly lower frequency of certain technology fields appearing in patents, in terms of the overall technology network, the connected technology nodes were more diverse and had an interdisciplinary nature in terms of applications. For example, although H04W4 and G06F19 are not listed in the top 10 technology fields with the highest frequency in Table 1, H04W4 was observed to connect to more different nodes in terms of degree centrality. Eigenvector centrality considered whether H04W4 was connected to a node with relatively high centrality, whereas structural holes revealed that G06F19 and H04W4 occupied the main channel of network communication; that is, the degree to which the connection between technology clusters depended on G06F1 and H04W4 was revealed.

4.3. Country-Technology Two-Mode Network Analysis

To include more interesting findings, country-technology two-mode network analysis and factionalization analysis were used to understand the strategic cluster of patent technology deployment of each country. Factions analysis was employed to conduct a complete survey of small-world structures. Factions analysis is an explorative tool used to identify subclusters in a social network [36]. In all, four factions were present. The final proposition correct was 0.703, suggesting a favorable fit value. The faction analysis results are presented in Table 6 and Figure 3.

Table 6. Faction analysis results.

Faction	Actors
Faction 1	G06F1, G06Q50, H04L9, H04N5, AU, BE
Faction 2	G06F17, G06K9, H04B10, H04L12, H04L29, CH, FI, GB, IL, IN, NL
Faction 3	A61B5, G06F9, G06F15, AT, CA, CN, DE
Faction 4	G06F3, G06F19, G06Q10, G06T7, G10L15, H04N7, H04W4, FR, IE, IT, JP, KR, LU, NO, PL, SG, US

Note: To simplify the table, only technology nodes that appear in Tables 1 and 5, and Figure 2 are presented.

Table 6 and Figure 3 can be used to identify the technology clusters of invention in the most prominent countries and the proximity of technical fields. For example, China and Germany belong to the same technology cluster, with their patents presenting high technical closeness in the A61B5 field. Australia and Belgium have high connectivity in the G06Q50 and H04L9 fields. Switzerland, the United Kingdom, and Israel have high connectivity in the G06F17, G06K9, and H04B10 fields. The United States and Japan have high technical closeness in the G06F3 and H04W4 fields.

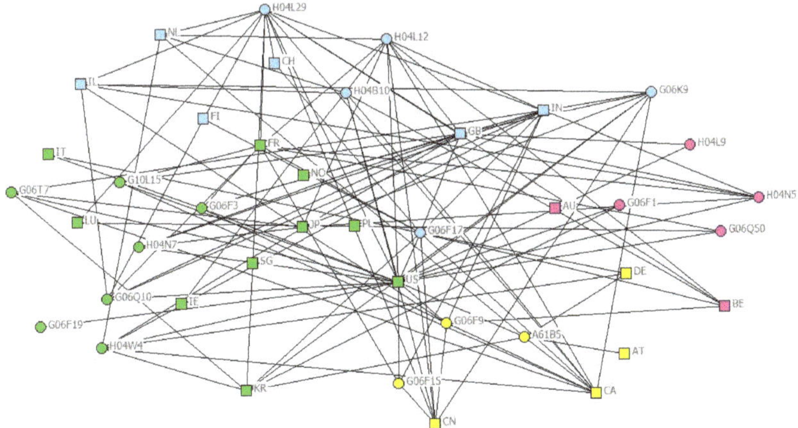

Figure 3. Two-mode network diagram based on faction analysis. Note: To simplify the figure, only technology nodes that appear in Tables 1 and 5, and Figure 2 are used.

4.4. Key Machine-Learning Technologies in Optical Communications over the Years

The changes in G06F3, G06F17, H04W4, H04L12 patents over the years were analyzed to understand the development trajectory of machine learning in optical communications (Figure 4).

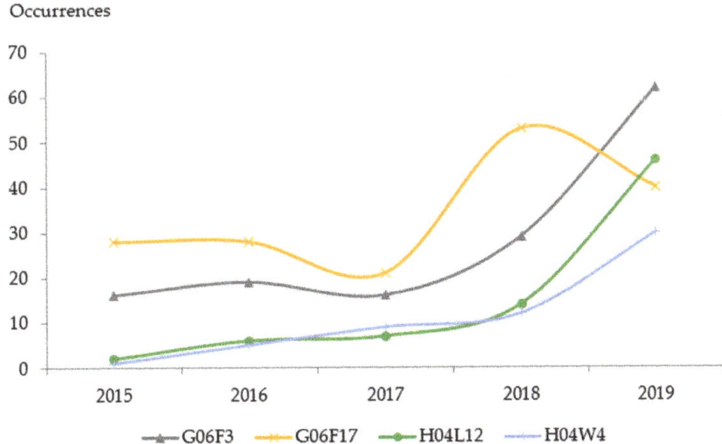

Figure 4. Key machine-learning technologies in optical communications over the years.

The results indicated that the application and development of G06F3 technologies—optical signals or data input and output devices—have gradually received more attention in recent years (Figure 2).

5. Conclusions

5.1. Results Discussion

In this study, network analysis was performed to explore key machine learning technologies in optical communications. The findings revealed that among data input and output devices, data-processing methods, wireless communication networks, and digital information transmission were key technologies that were not clustered in specific fields. The findings also indicated that machine-learning technologies in optical communications were applied to measurement for diagnostic purposes. Therefore, medical diagnostic applications is a direction that merits future study. In addition, comparing the differences between the network analysis and the most frequent Top 10 IPCs to the fourth digit revealed that wireless communication networks (H04W4) were among the top five in the network analysis; however, in the frequency analysis, their frequency of occurrence was not in the top 10. This indicated that although few patents for machine-learning technologies in optical communications were directly related to wireless communication networks, the connected technology fields were quite diverse. This highlights the importance of wireless communication networks and the advent of the era of wireless optical communications.

Furthermore, analysis of major patentees revealed that major developers are leaders in ICT and intelligent software and services—Apple Inc., SAS Institute Inc., Intel Corporation, and International Business Machines Corporation. Apple Inc. focuses on AI applications in communications, whereas SAS Institute Inc. conducts data exploration to advance algorithmic applications to deploy AI to more industries. Intel Corporation and International Business Machines Corporation are major vendors in chips and information. Although most of the top 10 patentees are leading ICT developers and the number of academic patents is lower than that of patents under general enterprises [37], academic and scientific research has a substantial influence on technology patents. The methods and specific techniques proposed in academic and scientific studies affect industrial development. For example, in the sample analyzed in this study, Carnegie Mellon University's patent (US10436615B2), authorized in 2019, uses machine learning. The computer system then trains a classifier to serve as a virtual sensor for an event that is correlated to the data from one or more sensor streams within the featured sensor data. The technology is related to the recording of measured values and can be applied in many industrial fields. Another example is a University of Central Oklahoma patent authorized in 2018 (US9922291B2) that proposes a method and apparatus for providing personalized configuration of physical supports for the human body, comprising accepting input including an individual's demographic information. The patents provide new methods and modes of thinking for computer systems based on specific computational models.

Factions analysis was adopted to determine the technical identification of inventions in prominent countries and to provide references for governments with regard to patent deployment. The results of the factions analysis revealed competition between countries in inventions, as well as the focal fields of each country. China and Germany belong to the same technical cluster, whereas the United States and Japan are in different clusters from those of the United Kingdom and Israel. Moreover, the proximity of technical fields can be observed through the results. For example, methods and devices for recognizing patterns (G06K9) and transmission systems employing electromagnetic waves other than radio-waves (H04B10) belong to the same faction; the co-occurrence of these technological fields in the same patent is highly likely. In technical applications of machine learning in optical communications, high co-occurrence and closeness are present.

The development trend of key machine-learning technologies in optical communications was concentrated in G06F3 or data input and output devices. Recent studies have argued that using machine learning and neural networks can precisely reconstruct digital images, convert blurry and unrecognizable speckle patterns into recognizable digital images, and process distortions caused by environmental disturbances to optic fibers [38]. This technological development is anticipated to advance endoscopic imaging and medical diagnoses [38,39]. In addition, in terms of causality of technology time series, patents for machine-learning technologies in optical communications were mainly focused on the technology field of data processing methods (G06F17), which emphasizes complex mathematical computation. Recently, optical signals as well as data input and output devices (G06F3) began to increase substantially. This indicates that with the development of communication technology and big data, in addition to the improvement of early mathematical computation through the development and application of technology, input devices that transform signals into digital data formats that can be processed by computers have gradually attracted attention, leading to further technological developments and relevant applications.

5.2. Implications Discussion

For theoretical contributions, studies on machine learning in optical communications have mostly explored algorithmic techniques [4,7] or researched specific applications [2,8]. However, these studies have failed to identify focal technical fields, development trends, and network distribution channels among technical fields from a macro perspective, particularly regarding indispensable technologies for the future development of machine learning for optical communications. This observation of technological distribution is particularly critical. This study filled this research gap and adopted a new perspective that centers on technical fields.

In technology development, as was described earlier, the focus of key machine-learning technologies in the optical communications field has been mainly on G06F3. In other words, machine learning is applied to data input and output devices in optical communications. Whether optical communications can achieve the prediction of transmission quality through machine learning is crucial in the development of optical communications technology. Traditional optical communications performance relies on the calculation of network layer parameters, which are based solely on the available flow and flow load of the network. However, whether optical communications are blocked is determined by not only the network layer but also the physical layer. The optical communications network must effectively predict the quality of transmission before new channel deployment of optical communications. The quality of transmission involves physical layer parameters such as signal-to-noise ratio and symbol error rate. How machine learning can be used to effectively predict the quality of transmission is the key to future technology development. The use of analysis models to estimate the damage of physical layers for the provision of accurate results is a fundamental challenge in the implementation of optical communications.

To assist in policy suggestions, this paper provides industries and governments with valuable information in a technical map of machine learning in optical communications. The analysis and understanding of technological development foci in technology networks inform industries on allocating research and development resources and informs governments on promoting emerging technologies. Given the development of IoT applications for driving the transmission volume of digital data, the optical communication field will require the integration of machine learning to construct adaptive smart optical communication system networks. This study found that major key technologies were concentrated in electric digital data processing, wireless communication networks, and the transmission of digital information. Therefore, governments should implement long-term financial support and training programs for talent in these technologies to increase the overall research and development capacity of the optical communication industry.

5.3. Limitations and Future Research Directions

First, this study used the patent keywords classifications that were organized by Derwent as the basis for patent screening. Although the Derwent database has several hundred experts who examine publicly available patent information to manually sort technical keywords, the development of machine-learning technologies in optical communications spans several fields and technical applications; therefore, some patents may not have been included in this analysis despite falling within its scope. Furthermore, this study analyzed large-scale holistic technology networks. Therefore, the empirical basis was limited to the number of approved patents, and the study did not make value judgements for individual patents. For example, this study lacked a discussion on whether the income of the patentees was correlated to the patents they owned, as well as the cost structure that patents imposed on the patentees. In the future, individual case studies into specific high-value patents can include expert interviews or other research methods. Finally, due to personnel and financial reasons, this study only used USPTO—the largest global commercial trading market—as its source of information on patents. Although this database is widely used to measure global innovations [13,14], future researchers with sufficient time and money should include other data sources on patents for observation and verification, such as by including approved standard essential patents in communications or their citation documents to expand their research breadth.

Funding: This research was funded by the Ministry of Science and Technology of the Taiwan, grant number MOST 109-2410-H-492-001.

Acknowledgments: The authors would like to thank the Ministry of Science and Technology of the Taiwan for financially supporting this research under Contract No. MOST 109-2410-H-492-001.

Conflicts of Interest: The authors declare no conflict of interest.

Appendix A

Table A1. Definition of IPC categories.

IPC Categories	Meaning
A61B5	Measuring for diagnostic purposes
G06F1	Details not covered by groups G06F 3/00-G06F 13/00 and G06F 21/00
G06F3	Input arrangements for transferring data to be processed into a form capable of being handled by the computer; Output arrangements for transferring data from processing unit to output unit, e.g., interface arrangements
G06F9	Arrangements for program control, e.g., control units
G06F15	Digital computers in general
G06F17	Digital computing or data processing equipment or methods, specially adapted for specific functions
G06F19	(transferred to G16C 10/00-G16C 60/00, G16Z 99/00); G16C 10/00: Computational theoretical chemistry, i.e., ICT specially adapted for theoretical aspects of quantum chemistry, molecular mechanics, molecular dynamics or the like; G16C 60/00: Computational materials science, i.e., ICT specially adapted for investigating the physical or chemical properties of materials or phenomena associated with their design, synthesis, processing, characterisation or utilization; G16Z 99/00: Subject matter not provided for in other main groups of this subclass
G06K9	Methods or arrangements for reading or recognising printed or written characters or for recognising patterns, e.g., fingerprints
G06N3	Computer systems based on biological models
G06N5	Computer systems using knowledge-based models
G06N99	Subject matter not provided for in other groups of this subclass
G06Q10	Administration; Management
G06Q50	Systems or methods specially adapted for specific business sectors, e.g., utilities or tourism
G06T7	Image analysis
G10L15	Speech recognition
H04B10	Transmission systems employing electromagnetic waves other than radio-waves, e.g., infrared, visible or ultraviolet light, or employing corpuscular radiation, e.g., quantum communication
H04L9	Arrangements for secret or secure communication
H04L12	Data switching networks
H04L29	Arrangements, apparatus, circuits or systems, not covered by a single one of groups
H04N5	Details of television systems
H04N7	Television systems
H04W4	Services specially adapted for wireless communication networks; Facilities thereof

Appendix B

Table A2. Centrality indicators of the Top 10 IPC codes with the highest frequency.

IPC	Degree Centrality	Eigenvector Centrality	Structural Hole
G06F17	102	0.151	0.085
G06F3	126	0.206	0.063
G06K9	109	0.139	0.090
H04L29	95	0.182	0.087
A61B5	91	0.112	0.076
G06F9	66	0.159	0.101
H04L12	97	0.176	0.088
G10L15	33	0.062	0.228
H04B10	90	0.169	0.085
G06T7	78	0.107	0.140

References

1. Liu, R.; Liu, Y.; Yan, Y.; Wang, J.Y. Iterative deep neighborhood: A deep learning model which involves both input data points and their neighbors. *Comput. Intell. Neurosci.* **2020**, 1–11. [CrossRef]
2. Ragheb, A.; Saif, W.; Trichili, A.; Ashry, I.; Esmail, M.A.; Altamimi, M.; Almaiman, A.; Altubaishi, E.; Ooi, B.S.; Alouini, M.S.; et al. Identifying structured light modes in a desert environment using machine learning algorithms. *Opt. Express* **2020**, *28*, 9753–9763. [CrossRef] [PubMed]
3. Turpin, A.; Vishniakou, I.; Seelig, J.D. Light scattering control in transmission and reflection with neural networks. *Opt. Express* **2018**, *26*, 30911–30929. [CrossRef]
4. Xu, Y.; He, D.; Wang, Q.; Guo, H.; Li, Q.; Xie, Z.; Huang, Y. An improved method of measuring wave front aberration based on image with machine learning in free space optical communication. *Sensors* **2019**, *19*, 3665. [CrossRef]
5. Argyris, A.; Bueno, J.; Fischer, I. Photonic machine learning implementation for signal recovery in optical communications. *Sci. Rep.* **2018**, *8*, 8487. [CrossRef] [PubMed]
6. Cui, X.; Yin, X.; Chang, H.; Liao, H.; Chen, X.; Xin, X.; Wang, Y. Experimental study of machine-learning-based orbital angular momentum shift keying decoders in optical underwater channels. *Opt. Commun.* **2019**, *452*, 116–123. [CrossRef]
7. Chugh, S.; Gulistan, A.; Ghosh, S.; Rahman, B.M.A. Machine learning approach for computing optical properties of a photonic crystal fiber. *Opt. Express* **2019**, *27*, 36414–36425. [CrossRef] [PubMed]
8. Zong, Y.; Feng, C.; Guan, Y.; Liu, Y.; Guo, L. Virtual network embedding for multi-domain heterogeneous converged optical networks: Issues and challenges. *Sensors* **2020**, *20*, 2655. [CrossRef] [PubMed]
9. Donges, A.; Selgert, F. Technology transfer via foreign patents in Germany, 1843–1877. *Econ. Hist. Rev.* **2019**, *72*, 182–208. [CrossRef]
10. Seo, I.; Sonn, J.W. The persistence of inter-regional hierarchy in technology transfer networks: An analysis of Chinese patent licensing data. *Growth Chang.* **2019**, *50*, 145–163. [CrossRef]
11. Aaldering, L.J.; Song, C.H. Tracing the technological development trajectory in post-lithium-ion battery technologies: A patent-based approach. *J. Clean. Prod.* **2019**, *241*, 118343. [CrossRef]
12. Gwak, J.H.; Sohn, S.Y. A novel approach to explore patent development paths for subfield technologies. *J. Assoc. Inf. Sci. Technol.* **2018**, *69*, 410–419. [CrossRef]
13. Kogler, D.F.; Heimeriks, G.; Leydesdorff, L. Patent portfolio analysis of cities: Statistics and maps of technological inventiveness. *Eur. Plan. Stud.* **2018**, *26*, 2256–2278. [CrossRef]
14. Yan, B.; Luo, J. Filtering patent maps for visualization of diversification paths of inventors and organizations. *J. Assoc. Inf. Sci. Technol.* **2017**, *68*, 1551–1563. [CrossRef]
15. Helmers, L.; Horn, F.; Biegler, F.; Oppermann, T.; Müller, K.-R. Automating the search for a patent's prior art with a full text similarity search. *PLoS ONE* **2019**, *14*, e0212103. [CrossRef]

16. Woo, H.-G.; Yeom, J.; Lee, C. Screening Early stage ideas in technology development processes: A text mining and K-nearest neighbours approach using patent information. *Technol. Anal. Strateg. Manag.* **2019**, *31*, 532–545. [CrossRef]

17. Cetintas, S.; Si, L. Effective query generation and postprocessing strategies for prior art patent search. *J. Am. Soc. Inf. Sci. Technol.* **2012**, *63*, 512–527. [CrossRef]

18. Hansen, P.; Järvelin, A.; Järvelin, A. Exploring manual and automatic query formulation in patent IRInitial query construction and query generation process. *J. Doc.* **2013**, *69*, 873–898. [CrossRef]

19. Mahdabi, P.; Crestani, F. The effect of citation analysis on query expansion for patent retrieval. *Inf. Retr. J.* **2014**, *17*, 412–429. [CrossRef]

20. MarketsandMarkets. *Optical Communication and Networking Equipment Market by Component (Fiber, Transceiver, and Switch), Technology, Application (Telecom, Data Center, and Enterprise), Data Rate, Vertical, and Region–Global Forecast to 2025*; MarketsandMarkets: Northbrook, IL, USA, 2020.

21. Mata, J.; de Miguel, I.; Durán, R.J.; Merayo, N.; Singh, S.K.; Jukan, A.; Chamania, M. Artificial intelligence (AI) methods in optical networks: A comprehensive survey. *Opt. Switch. Netw.* **2018**, *28*, 43–57. [CrossRef]

22. Gu, R.; Yang, Z.; Ji, Y. Machine learning for intelligent optical networks: A comprehensive survey. *J. Netw. Comput. Appl.* **2020**, *157*, 102576. [CrossRef]

23. Khan, F.N.; Fan, Q.; Lau, A.P.T.; Lu, C. Applications of machine-learning in optical communications and networks. Next-Generation Optical Communication: Components, Sub-Systems, and Systems IX. In Proceedings of the 2020 Optical Fiber Communications Conference and Exhibition (OFC), San Diego, CA, USA, 8–12 March 2020; p. 113090L.

24. Cochoy, F.; Soutjis, B. Back to the future of digital price display: Analyzing patents and other archives to understand contemporary market innovations. *Soc. Stud. Sci.* **2020**, *50*, 3–29. [CrossRef] [PubMed]

25. Kim, K.; Jung, S.; Hwang, J. Technology convergence capability and firm innovation in the manufacturing sector: An approach based on patent network analysis. *RD Manag.* **2019**, *49*, 595–606. [CrossRef]

26. de Paulo, A.F.; Ribeiro, E.M.S.; Porto, G.S. Mapping countries cooperation networks in photovoltaic technology development based on patent analysis. *Scientometrics* **2018**, *117*, 667–686. [CrossRef]

27. Zhu, L.; Zhu, D.; Wang, X.; Cunningham, S.W.; Wang, Z. An integrated solution for detecting rising technology stars in co-inventor networks. *Scientometrics* **2019**, *121*, 137–172. [CrossRef]

28. Chen, Z.; Guan, J. The core-peripheral structure of international knowledge flows: Evidence from patent citation data. *RD Manag.* **2016**, *46*, 62–79. [CrossRef]

29. Sonmez, Z. Inventor mobility and the geography of knowledge flows: Evidence from the US biopharmaceutical industry. *Sci. Public Policy* **2017**, *44*, 670–682. [CrossRef]

30. Kwon, O.; An, Y.; Kim, M.; Lee, C. Anticipating technology-driven industry convergence: Evidence from large-scale patent analysis. *Technol. Anal. Strateg. Manag.* **2020**, *32*, 363–378. [CrossRef]

31. Mun, C.; Yoon, S.; Park, H. Structural decomposition of technological domain using patent co-classification and classification hierarchy. *Scientometrics* **2019**, *121*, 633–652. [CrossRef]

32. Leydesdorff, L.; Nooy, W. Can 'hot spots' in the sciences be mapped using the dynamics of aggregated journal-journal citation Relations? *J. Assoc. Inf. Sci. Technol.* **2017**, *68*, 197–213. [CrossRef]

33. Borgatti, S.P.; Everett, M.G.A. Graph-theoretic perspective on centrality. *Soc. Netw.* **2006**, *28*, 466–484. [CrossRef]

34. Burt, R.S. *Structural Holes*; Harvard University Press: Cambridge, MA, USA, 1992.

35. Tseng, C.Y.; Ting, P.H. Patent analysis for technology development of artificial intelligence: A country-level comparative study. *Innov. Manag. Policy Pract.* **2013**, *15*, 463–475. [CrossRef]

36. Cunningham, D.; Everton, S.; Murphy, P. *Understanding Dark Networks: A Strategic Framework for the Use of Social Network Analysis*; Rowman & Littlefield: Lanham, MD, USA.

37. Wong, P.K.; Singh, A. University patenting activities and their link to the quantity and quality of scientific publications. *Scientometrics* **2010**, *83*, 271–294. [CrossRef]

38. Borhani, N.; Kakkava, E.; Moser, C.; Psaltis, D. Learning to see through multimode fibers. *Optica* **2018**, *5*, 960–966. [CrossRef]

39. Flusberg, B.A.; Cocker, E.D.; Piyawattanametha, W.; Jung, J.C.; Cheung, E.L.M.; Schnitzer, M.J. Fiber-optic fluorescence imaging. *Nat. Methods* **2005**, *2*, 941–950. [CrossRef]

Publisher's Note: MDPI stays neutral with regard to jurisdictional claims in published maps and institutional affiliations.

 *photonics*

Article

Comparing Performance of Deep Convolution Networks in Reconstructing Soliton Molecules Dynamics from Real-Time Spectral Interference

Caiyun Li , Jiangyong He, Yange Liu, Yang Yue, Luhe Zhang, Longfei Zhu, Mengjie Zhou, Congcong Liu, Kaiyan Zhu and Zhi Wang *

Tianjin Key Laboratory of Optoelectronic Sensor and Sensing Network Technology, Institute of Modern Optics, Nankai University, Tianjin 300350, China; 1120180099@mail.nankai.edu.cn (C.L.); 1120190097@mail.nankai.edu.cn (J.H.); ygliu@nankai.edu.cn (Y.L.); yueyang@nankai.edu.cn (Y.Y.); 2120180269@mail.nankai.edu.cn (L.Z.); 2120180259@mail.nankai.edu.cn (L.Z.); 2120190246@mail.nankai.edu.cn (M.Z.); 1120200110@mail.nankai.edu.cn (C.L.); 2120190247@mail.nankai.edu.cn (K.Z.)
* Correspondence: zhiwang@nankai.edu.cn

Citation: Li, C.; He, J.; Liu, Y.; Yue, Y.; Zhang, L.; Zhu, L.; Zhou, M.; Liu, C.; Zhu, K.; Wang, Z. Comparing Performance of Deep Convolution Networks in Reconstructing Soliton Molecules Dynamics from Real-Time Spectral Interference. *Photonics* **2021**, *8*, 51. https://doi.org/10.3390/photonics8020051

Academic Editor: Yuri Kivshar

Received: 30 December 2020
Accepted: 9 February 2021
Published: 13 February 2021

Publisher's Note: MDPI stays neutral with regard to jurisdictional claims in published maps and institutional affiliations.

Abstract: Deep neural networks have enabled the reconstruction of optical soliton molecules with more complex structures using the real-time spectral interferences obtained by photonic time-stretch dispersive Fourier transformation (TS-DFT) technology. In this paper, we propose to use three kinds of deep convolution networks (DCNs), including VGG, ResNets, and DenseNets, for revealing internal dynamics evolution of soliton molecules based on the real-time spectral interferences. When analyzing soliton molecules with equidistant composite structures, all three models are effective. The DenseNets with layers of 48 perform the best for extracting the dynamic information of complex five-soliton molecules from TS-DFT data. The mean Pearson correlation coefficient (MPCC) between the predicted results and the real results is about 0.9975. Further, the ResNets in which the MPCC achieves 0.9906 also has the better ability of phase extraction than VGG which the MPCC is about 0.9739. The general applicability is demonstrated for extracting internal information from complex soliton molecule structures with high accuracy. The presented DCNs-based techniques can be employed to explore undiscovered mechanisms underlying the distribution and evolution of large numbers of solitons in dissipative systems in experimental research.

Keywords: fiber nonlinearities; deep learning (DL); artificial intelligence (AI)

1. Introduction

Soliton molecules are localized soliton bound states formed by self-organized dissipative soliton through subtle interaction mechanisms [1]. The potential of soliton molecules to expand the transmission capacity in optical communication systems has drawn much research attention and has become an attractive topic for nonlinear optical fibers in recent decades [2–10]. In addition to predicting the dynamic evolution of soliton molecules theoretically [3], the dynamic evolution of soliton molecules is also proved experimentally [8–12], which extends the degrees of freedom toward internal dynamics. The internal dynamics of soliton molecules is difficult to analyze when only the change of the pulse energy is considered in the oscilloscope traces. Recently, the photonic time-stretch dispersive Fourier transformation (TS-DFT) technology has been used to real-time monitor the internal dynamics of soliton molecules in passive mode-locked lasers (PMLs). Concretely, TS-DFT observe various rare events and transient phenomena including soliton buildup [6,7], soliton pulsation [13,14], soliton explosion [15,16], and soliton molecules [2–4,9,10]. It appears tremendous potential in simulating dynamic process of various complex molecules. The structure of soliton molecules encompasses simple two-soliton and three-soliton molecules [4,5,10–12], 2+2 soliton molecular complexes [9], composite patterns in both global and local ranges [14], and supramolecular arrangements that mimic various many-body biochemical and biological systems [8]. To restructure the internal dynamics of the soliton molecules from TS-DFT spectra, a autocorrelation method is usually employed [9]. In this method, a discrete Fourier transform is performed on the interference fringes to obtain the single-shot autocorrelation traces for retrieving the soliton separation and the relative phase in the soliton molecules.

However, the autocorrelation method cannot further quantitatively analyze all the dynamic evolution processes in complex molecular structures, such as relative phase of each soliton. When multisoliton molecules and soliton pairs with near equal spacing happen [11], it is almost impossible to obtain relative phase differences (PDs) evolution [17]. Therefore, the autocorrelation method is suitable for analyzing simple soliton molecule structures consisting of soliton pairs, unequally spacing three solitons [4], etc. Recent years have seen the rapid growth and development of the field of ultrafast photonics, where artificial intelligence algorithms are being applied in exploring complex dynamical processes of soliton molecules in PMLFLs [17], the extreme events in optical fibre modulation instability [18], and the generation and characterization of light pulses [19,20]. In order to solve the internal dynamics of complex soliton molecules, we introduced artificial intelligence combining with TS-DFT. Although the residual networks (ResNets) [21] have been used for exploring complex dynamical processes in soliton molecules experimentally and numerically based on TS-DFT in passive mode-locked lasers (PMLs), emerging models continue to push the limits of what can be achieved. It also has proved that the data generated based on theory can be used to analyze experimental data [17]. It is necessary to consider whether the network structures outside the ResNets are more accurate and effective to analyze the internal dynamics of multisoliton molecules.

Recently, deep convolution networks (DCNs) have demonstrated a powerful ability to apply in mode-locked lasers [17,22], decompose the modes in few-mode fibers [23], recognize orbital angular momentum modes with fractional topological charges [24], mitigate fiber nonlinearity in optical communication [25], and the characterization and control of ultrafast propagation dynamics [26]. It is well known that convolutional neural networks (CNNs) have dominated machine-learning landscape in data-rich applications, such as VGG (Visual Geometry Group) [27], Residual Networks (ResNets) [21], Dense Convolutional Networks (DenseNets) [28], and other models. Theoretical and empirical evidences indicate that the depth of neural networks is crucial for its accuracy and/or performance [29]. The core of DenseNets and ResNets models is to establish "shortcuts, Skip Connection" between the front and back layers, which will facilitate shortcuts and skip connections during training and enable deeper CNN networks to be trained and achieve higher accuracy. The difference in DenseNets model is that each layer can directly obtain the gradients from the loss function and the original input signal, thus forming an implicit form of deep supervision [30,31]. This makes the feature reuse through the connection of features across the channel for faster error converge. Considering the representativeness of VGG, ResNets, and DenseNets models and their characteristics of easily deepening the network layers, the three kinds of models are chosen to compare the ability in extracting internal dynamics evolution of soliton molecules.

Here, we propose and demonstrate, theoretically, the analysis the internal dynamics of bound states of complex dissipative solitons by employing DCNs. We implement VGG, ResNets, and DenseNets which are able to extract the phase evolution information of more complex soliton molecules from TS-DFT spectra data by modifying the network structure. Comparing the performance of the three DCNs by numerical analysis, the ResNets and DenseNets represent lower complexity than VGG and can easily enjoy accuracy gains from greatly increased layers. The DenseNets we used have better parameter efficiency and more lower error than ResNets in the test data. Thus, DenseNets have been demonstrated to achieve superior performance in comparison to other two models by almost any meaningful metric.

2. Methods

2.1. Generate Simulated TS-DFT Data of Soliton Molecules

The generation of simulated TS-DFT data of soliton bound state is considering factors such as bandwidth, sampling, and noise, which has been proven to be used for deep learning data sets [17]. The complex amplitude of the slowly varying envelope of soliton molecules is described by the superposition of solitons, which is given by [32]

$$U_M(T) = \sum_{k=1}^{M} u_k(T - \tau_k)e_k^{-i\varphi_k},\tag{1}$$

where T is the relative reference time of the pulse, M is the number of solitons, and u_k, τ_k, and φ_k represent the slowly varying envelope, relative temporal delay, and relative phase of the k-th soliton, respectively. When the bandwidth and sampling speed of the electronic devices are matching with experiment, for example, the parameters of a real-time oscilloscope are 59 GHz and 200 GSa/s, the TS-DFT spectrum with resolution of 2.8626 ps is calculated first with high temporal resolution (0.01 ps) and then filtered by a fourth-order Butterworth lowpass filter and downsampled. Thus, the simulated TS-DFT dataset for the soliton molecules can be acquired based on a series of relative temporal delay τ_k and phase φ_k is given. All the TS-DFT data are superimposed white noise. The TS-DFT system, we used here, has a dispersion-compensating fiber (DCF, -134 ps2/km) with length of 1.5 km. We assume that the solitons in soliton molecules are hyperbolic secant pulses with a central wavelength of 1560 nm. As shown in Figure 1, when multisoliton molecules are considered, the TS-DFT dataset is generated with random PDs. The TS-DFT dataset is filtered and divided into a training set and a verification set proportionally (8:2). All the TS-DFTs are converted to bitmap for the inputs of DCNs. After the training via DCNs, the simulated testing dataset, with noise, is used to predict the PDs of the soliton dynamics.

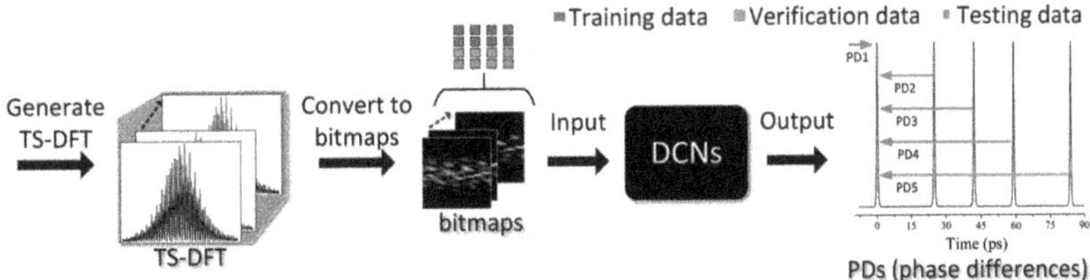

Figure 1. Processing flow for TS-DFT data for the soliton molecules based on the neural networks.

2.2. Structures of Deep Convolution Networks (DCNs)

The based architectures of three DCNs, namely VGG, ResNets, and DenseNets, are ref. [28,33,34]. We made some modification in these three models, including the number of layers of the network, the size of convolution kernel, and the structure of subblock. Especially, in Figure 2a, a batch normalization (BN) is added before each convolution block unlike VGG nets in ref [35]. Meanwhile a regularization $L2$-norm is used in each convolutional layer. The convolutional layers have the same convolution kernel (K_i) in one convolution block. With the stack of convolution blocks, the number of convolution kernels increases or is the same as the previous block. The main parts of ResNets/DenseNets are made up of their ResBlocks/Dense Blocks as shown in Figure 2b,c. The number of subblocks for each ResBlock/Dense Block is set respectively. Their structure of the subblocks are displayed in the box pointed to by the arrow. In addition, all the convolutional layers with regularization $L2$-norm are employed and batch normalization is applied among the layers. The activation function, which uses the rectified linear unit (ReLU) [36] and the Batch-Normalization [37], regularization $L2$-norm, and pooling, used in our three DCNs, can prevent overfitting. The regularization $L2$-norm makes the objective function easy to converge to the global optimal solution. The weights of the DCNs are optimized during the training process through backpropagation. The optimizer we used is Adam [38], a variant of stochastic gradient descent that has individual adaptive learning rates for different parameters, which are calculated from estimates of the first and second moments of the gradients. Moreover, the mean absolute error (MAE) is chosen here because DCNs implement regression problems. The function of the optimizer is to reduce the gap between the predicted value and the sample label value. The DCNs' models are implemented using the Tensorflow framework [39].

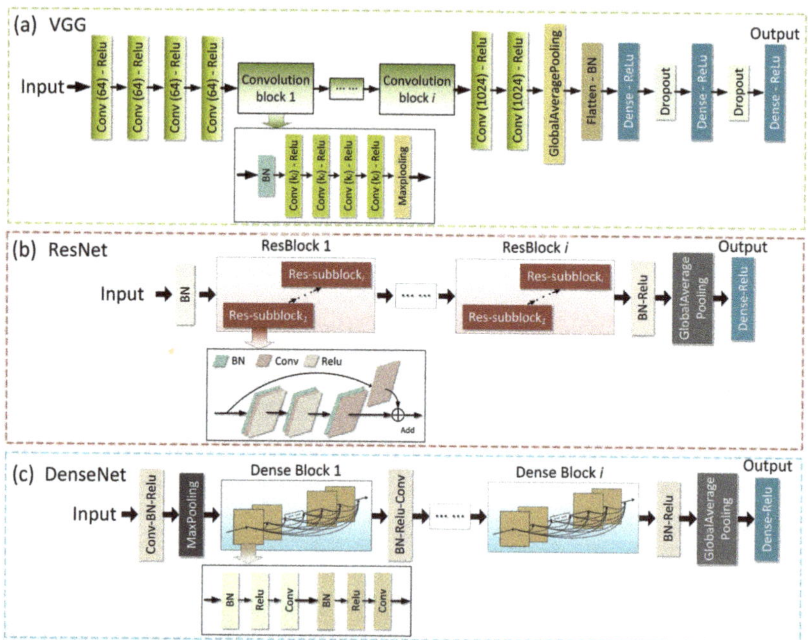

Figure 2. The structure of the DCNs (deep convolution networks). (**a**) The VGG Net (Visual Geometry Group Networks). (**b**) The ResNet (Residual Networks). (**c**) The DenseNet (Densely Connected Convolutional Networks).

3. Results and Discussion

3.1. Soliton Molecular Structure of Test Set

A complex soliton molecular structure with five solitons, which is exhibited in Figure 3, is used to test the ability of the three DCNs in extracting relative phase differences (PDs). In particular, the internal phase evolution of the soliton molecules contain oscillating and the diverging sliding phase [4,5]. The test set includes both phases and the equal temporal separations so it is impossible to extract internal phase evolution of each soliton by autocorrelation method. The temporal trace of simulated dataset is shown in Figure 3a. The temporal separations of the five solitons contain two kinds of equal spacing 17 and 42 picoseconds (ps). As presented in Figure 3a, a phasor representation is constructed to picture the five-soliton molecules constituted. We defined the leftmost soliton as the first pulse which is set as the reference with a fixed pointing direction. Then, the PD from the following pulse to the first pulse are defined as PD2, PD3, etc., denoted by the variables (φ). Figure 3e lists two PDs as representatives containing oscillating and the diverging sliding phase [40]. The TS-DFT of five-soliton molecules with given phases as the simulated testing dataset show in Figure 3b. Because there are soliton pairs with almost equal separation within the soliton molecule, their corresponding autocorrelation peaks are coherently superposed. The autocorrelation trajectories are flickering as shown in Figure 3c. Specifically, two roundtrips (580 and 704 roundtrips) of autocorrelation curves are drawn in Figure 3d. It is obvious that the intensity varies greatly at the autocorrelation peaks for the interaction of isometric soliton molecules. This complex molecular structure as a test set involves the difficulties mentioned above and has the ability to evaluate the merits and demerits of the DCNs.

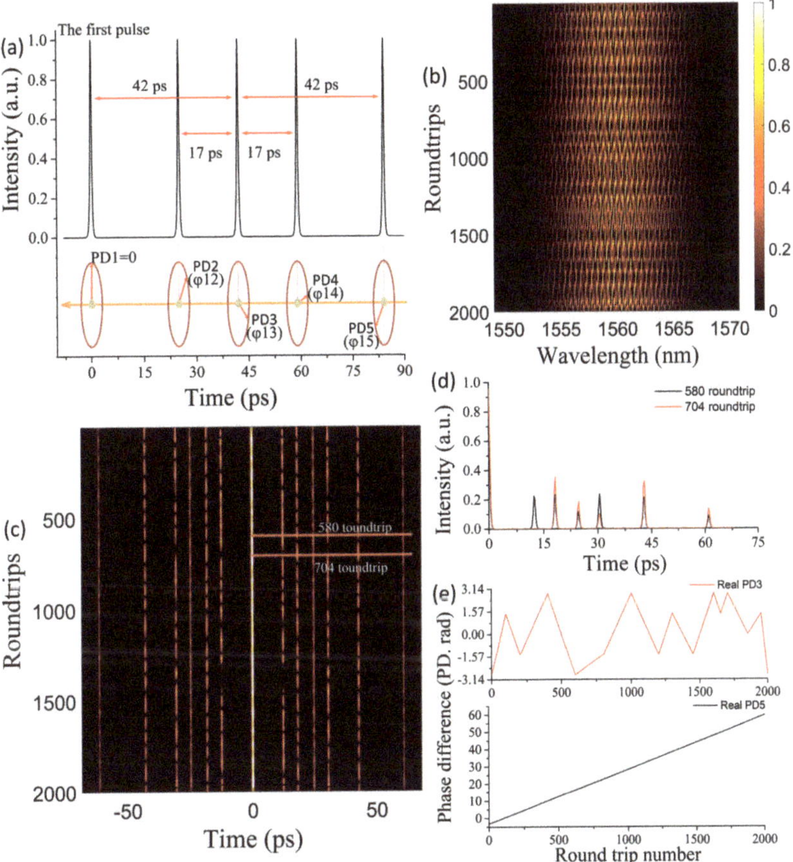

Figure 3. (**a**) Graphical representation of five-soliton molecules. (**b**) The testing TS-DFT (time-stretch dispersive Fourier transformation). (**c**) The autocorrelation trajectories of TS-DFT. (**d**) Comparison of two autocorrelation trajectories. (**e**) Two kinds of phase differences (PDs) evolution as representatives containing oscillating and the diverging sliding phases corresponding Real PD3 and PD5.

3.2. Perform Three DCNs on TS-DFT Datasets of Five-Soliton Molecules

The TS-DFT dataset, with 39×39 pixels each, put into three DCNs for training. We add three callback functions to control the program. These include dynamic adjustment of learning rate (LR) which is multiplied by 0.6 to decrease value if the error of lose function does not decrease after 5 iterations. The Early-Stop function is to terminate the program when the error of lose function does not decrease after 20 iterations. The Best-Model function saves optimal parameter model when the error is less than previous error. The training results are shown in Figure 4. The convergence speed and error of different networks are diverse because of the number of layers. As shown in Figure 4a, we considered for VGG of four network layers of 17, 21, 25, and 29. ResNet of three network layers of 65 (k = 512), 77 (k = 515), and 65 (k = 1024) are in Figure 4b. DenseNet of four network layers of 121 (k = 32), 161 (k = 32), 161 (k = 48), 169 (k = 32) and 169 (k = 48) are in Figure 4c.

Table 1 lists the depth of networks, the size of parameter model, the number of iterations, the verification errors and test errors of different model structures for TS-DFT of five-soliton molecules. Thereinto, the DenseNet of 161($k = 48$) has the best testing results with smallest error 2.2355 and faster convergence rate on the comprehensive. Because overfitting cannot be avoided completely and different networks have different inhibitory overfitting effects. Thus, the trends of verification error and testing error have a little inconsistency. From Figure 4, the error trend remains the same: the lower the verification error, the lower the testing error. Here we evaluate the accuracy of the networks mainly based on the error of the test data. It can be seen from Table 1 and Figure 4 that VGG networks have the worst effect for phase extraction. Its minimum testing error is high, 5.2528. DenseNet, with minimum testing error 2.2355, has a slightly smaller advantage over ResNet whose value is 2.6260. By comparing the verification errors of the optimal results in each DCNs, as shown in Figure 4d, we can still conclude that the VGG shows the worst convergence and the optimal one is the DenseNet, where the networks with shortcut connection can suppress gradient explosion better than the common convolutional network.

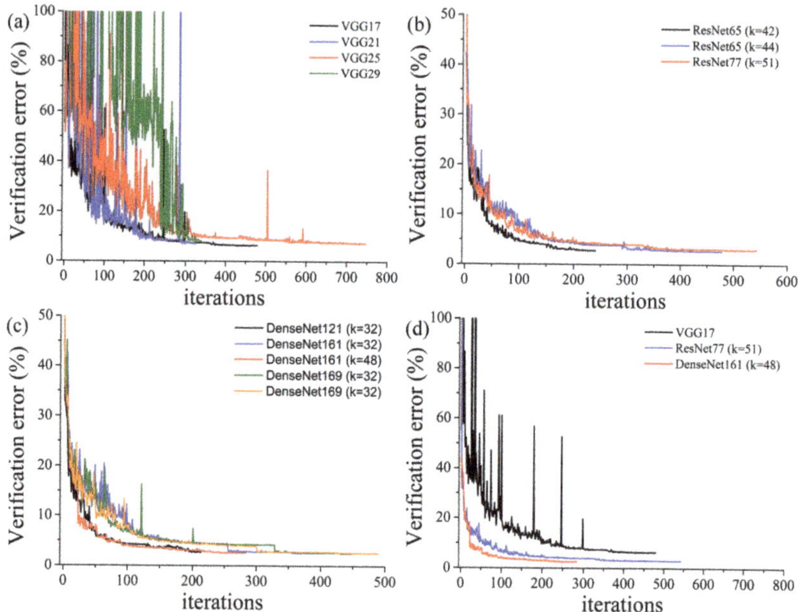

Figure 4. Verification error of DCNs. (**a**) Four VGG networks. (**b**) Three ResNets. (**c**) Five DenseNets. (**d**) Three DCNs with optimal test results.

3.3. Pearson Correlation Analysis of Real and Predicted Values

Next, we compare the real relative PDs (black lines) with the extraction results (red lines) from the optimal model in each DCN. The left column in Figure 5a is the PDs extracted from VGG-17 with a minimum error of 5.2528. Figure 5b plot the PDs extracted from ResNet-77 with a minimum error of 2.6260. In addition, the PDs extracted from DenseNet-161 ($k = 48$) with a minimum error of 2.2355 in Figure 5c. The correlation between the real value and the extracted value is analyzed by Pearson Correlation Coefficient (PCC). The mean Pearson correlation coefficients (MPCC) of each group of PDs are 0.9739, 0.9906, 0.9975 which correspond to DCNs of VGG-PDs, ResNet-PDs, and DenseNet-PDs, respectively. After comparing the VGG, ResNet, and DenseNet, the ResNets and DenseNet represent fewer smaller error and lower complexity than VGG and can easily enjoy accuracy gains from greatly increased layers. It is worth noticing that extremely deep nets with shortcut paths are easy to optimize, but simply stack layers exhibit higher testing error when the depth increases [21]. Because short paths in the network have a strong regularizing effect and reduce overfitting on smaller training sets [30]. Besides, DenseNets we used have better parameter efficiency and more lower error than ResNets in the test data. It has been reported that DenseNets are easier to train due to their improved information flow and gradients throughout the network [30,31]. On these, the DenseNets have the best testing results with smallest testing error and superior parameter efficiency on the comprehensive. They tend to require far fewer parameters when compared against alternative algorithms with comparable accuracy. Consequently, we infer that the DCNs model have the potential to analyze the dynamics of more complex soliton molecules and DenseNets performs best.

Table 1. Error rates (%) of single-model results on the TS-DFT interference spectra of five-soliton molecules datasets.

Model-Layers	Params	Iterations	Verification Error (%)	Testing Error (%)
VGG17	268 M	479	**6.2891**	**5.2528**
VGG21	272 M	324	7.1810	6.5479
VGG25	320 M	747	7.3101	6.8600
VGG29	332 M	339	8.1265	7.3815
ResNet65 ($k = 42$)	122 M	241	2.7159	2.9438
ResNet65 ($k = 44$)	426 M	543	**2.6445**	2.8491
ResNet77 ($k = 51$)	187 M	478	2.9573	**2.6260**
DenseNet121 ($k = 32$)	68.1 M	213	2.6361	2.6155
DenseNet161 ($k = 32$)	112 M	405	2.6057	2.5037
DenseNet161 ($k = 48$)	246 M	284	2.5917	**2.2355**
DenseNet169 ($k = 32$)	126 M	448	**2.5088**	2.7286
DenseNet169 ($k = 48$)	278 M	490	2.6103	2.8331

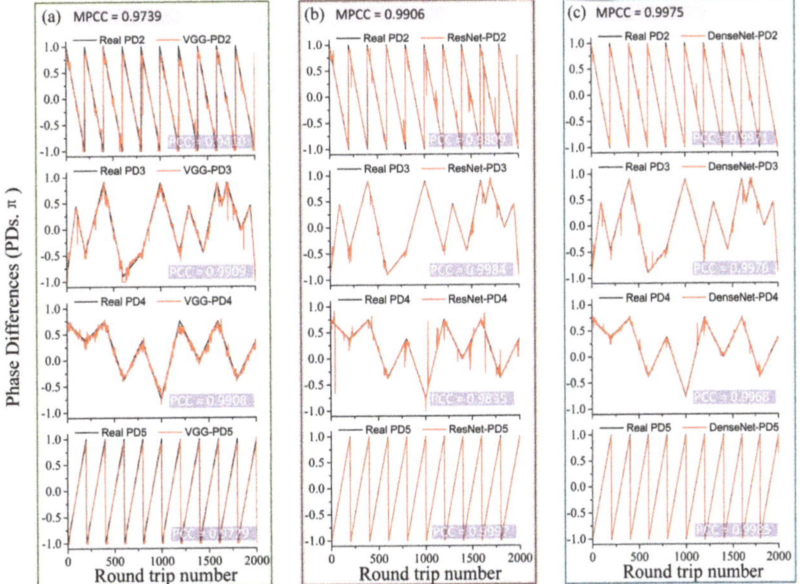

Figure 5. Relative phase difference (PD2-PD5). (**a**) The real PDs and VGG-PDs. (**b**) The real PDs and ResNet-PDs. (**c**) The real PDs and DenseNet-PDs.

4. Conclusions

The methods based on DCNs can solve the situation of more solitons and existence of equidistant soliton pairs where the autocorrelation method is limited. Comparing the

VGG, ResNet, and DenseNet models, we demonstrate their effectiveness on TS-DFT interference spectra of more complex five-soliton molecules datasets with equal spacing pairs. The DenseNets outperform VGG and ResNets in extracting the internal information from complex five-soliton molecules, where the second best is the ResNets whether considering parameter efficiency or testing error. The investigation on the soliton molecule in the PMLs would contribute to understanding the complex nonlinear dynamics of pulse propagation in PMLs and benefit the potential applications of telecommunications and fiber laser sources. This provides the possibility of simulating the dynamic behaviors of complex chemical molecules and other multibody systems based on soliton molecules in PMLs optically. We expect that our method can promote simulating the dynamic behaviors of complex chemical molecules and other multibody systems based on soliton molecules in PMLs optically and explore the potential mechanism of the distribution and evolution of a large numbers of solitons in a dissipative system.

Author Contributions: Conceptualization, C.L. (Caiyun Li), J.H. and Z.W.; methodology, C.L. (Caiyun Li); software, C.L. (Caiyun Li); validation, L.Z. (Luhe Zhang), L.Z. (Longfei Zhu), M.Z., C.L. (Congcong Liu) and K.Z.; formal analysis, C.L. (Caiyun Li) and Z.W.; investigation, C.L. (Congcong Liu); resources, Z.W., Y.Y. and Y.L.; data curation, C.L. (Caiyun Li); writing—original draft preparation, C.L. (Caiyun Li) and Z.W.; writing—review and editing, C.L. (Caiyun Li); visualization, C.L. (Caiyun Li) and Z.W.; supervision, Z.W.; project administration, Z.W., Y.Y. and Y.L.; funding acquisition, Z.W., Y.Y. and Y.L. All authors have read and agreed to the published version of the manuscript.

Funding: This work was jointly supported by the National Key Research and Development Program of China under Grant 2018YFB0504400, National Natural Science Foundation of China (NSFC) (61775107, 11674177, 61640408); Tianjin Natural Science Foundation (19JCZDJC31200), China.

Data Availability Statement: The data that support the plots within this paper and other findings of this study are available from the corresponding author upon reasonable request. The data processing and simulation codes that were used to generate the plots within this paper and other findings of this study are available from the corresponding author upon reasonable request.

Conflicts of Interest: This manuscript has not been published or presented elsewhere in part or in entirety and is not under consideration by another journal. We have read and understood your journal's policies, and we believe that neither the manuscript nor the study violates any of these. There are no conflicts of interest to declare.

Abbreviations

The following abbreviations are used in this manuscript:

TS-DFT	time-stretch dispersive Fourier transformation
DCNs	deep convolution networks
MPCC	mean Pearson correlation coefficient
PDs	relative phase differences

References

1. Grelu, P.; Akhmediev, N. Dissipative solitons for mode-locked lasers. *Nat. Photonics* **2012**, *6*, 84–92. [CrossRef]
2. Stratmann, M.; Pagel, T.; Mitschke, F. Experimental observation of temporal soliton molecules. *Phys. Rev. Lett.* **2005**, *95*, 143902. [CrossRef] [PubMed]
3. Zavyalov, A.; Iliew, R.; Egorov, O.; Lederer, F. Dissipative soliton molecules with independently evolving or flipping phases in mode-locked fiber lasers. *Phys. Rev. A* **2009**, *80*, 043829. [CrossRef]
4. Herink, G.; Kurtz, F.; Jalali, B.; Solli, D.R.; Ropers, C. Real-time spectral interferometry probes the internal dynamics of femtosecond soliton molecules. *Science* **2017**, *356*, 50–53. [CrossRef] [PubMed]
5. Krupa, K.; Nithyanandan, K.; Andral, U.; Tchofo-Dinda, P.; Grelu, P. Real-Time Observation of Internal Motion within Ultrafast Dissipative Optical Soliton Molecules. *Phy. Rev. Lett.* **2017**, *118*, 243901. [CrossRef] [PubMed]
6. Liu, X.; Yao, X.; Cui, Y. Real-Time Observation of the Buildup of Soliton Molecules. *Phy. Rev. Lett.* **2018**, *121*, 023905. [CrossRef] [PubMed]
7. Peng, J.; Zeng, H. Build-Up of Dissipative Optical Soliton Molecules via Diverse Soliton Interactions. *Laser Photonics Rev.* **2018**, *12*, 1800009. [CrossRef]

8. He, W.; Pang, M.; Yeh, D.H.; Huang, J.; Menyuk, C.R.; Russell, P.S.J. Formation of optical supramolecular structures in a fibre laser by tailoring long-range soliton interactions. *Nat. Commun.* **2019**, *10*, 5756. [CrossRef] [PubMed]

9. Wang, Z.Q.; Nithyanandan, K.; Coillet, A.; Tchofo-Dinda, P.; Grelu, P. Optical soliton molecular complexes in a passively mode-locked fibre laser. *Nat. Commun.* **2019**, *10*, 830. [CrossRef]

10. Kurtz, F.; Ropers, C.; Herink, G. Resonant excitation and all-optical switching of femtosecond soliton molecules. *Nat. Photonics* **2020**, *14*, 9–13. [CrossRef]

11. Luo, Y.; Xia, R.; Shum, P.; Ni, W.; Ys, L.; Lam, H.; Sun, Q.; Tang, X.; Zhao, L. Real-time dynamics of soliton triplets in fiber lasers. *Photonics Res.* **2020**, *8*, 884–891. [CrossRef]

12. Liang, H.; Zhao, X.; Liu, B.; Yu, J.; Liu, Y.; He, R.; He, J.; Li, H.; Wang, Z. Real-time dynamics of soliton collision in a bound-state soliton fiber laser. *Nanophotonics* **2020**, *9*, 1921–1929. [CrossRef]

13. Peng, J.; Boscolo, S.; Zhao, Z.; Zeng, H. Breathing dissipative solitons in mode-locked fiber lasers. *Sci. Adv.* **2019**, *5*, eaax1110. [CrossRef]

14. Wang, Z.; Wang, Z.; Liu, Y.; He, R.; Zhao, J.; Wang, G.; Yang, G. Self-organized compound pattern and pulsation of dissipative solitons in a passively mode-locked fiber laser. *Opt. Lett.* **2018**, *43*, 478–481. [CrossRef]

15. Runge, A.F.J.; Broderick, N.G.R.; Erkintalo, M. Observation of soliton explosions in a passively mode-locked fiber laser. *Optica* **2015**, *2*, 36–39. [CrossRef]

16. Wang, X.; Liu, Y.G.; Wang, Z.; Yue, Y.; He, J.; Mao, B.; He, R.; Hu, J. Transient behaviors of pure soliton pulsations and soliton explosion in an L-band normal-dispersion mode-locked fiber laser. *Opt. Express* **2019**, *27*, 17729–17742. [CrossRef] [PubMed]

17. Li, C.; He, J.; He, R.; Liu, Y.; Yue, Y.; Liu, W.; Zhang, L.; Zhu, L.; Zhou, M.; Zhu, K.; et al. Analysis of real-time spectral interference using a deep neural network to reconstruct multi-soliton dynamics in mode-locked lasers. *APL Photonics* **2020**, *5*, 116101. [CrossRef]

18. Närhi, M.; Salmela, L.; Toivonen, J.; Billet, C.; Dudley, J.M.; Genty, G. Machine learning analysis of extreme events in optical fibre modulation instability. *Nat. Commun.* **2018**, *9*, 4923. [CrossRef]

19. Boscolo, S.; Finot, C. Artificial neural networks for nonlinear pulse shaping in optical fibers. *Opt. Laser Technol.* **2020**, *131*, 106439. [CrossRef]

20. Kokhanovskiy, A.; Bednyakova, A.; Kuprikov, E.; Ivanenko, A.; Dyatlov, M.; Lotkov, D.; Kobtsev, S.; Turitsyn, S. Machine learning-based pulse characterization in figure-eight mode-locked lasers. *Opt. Lett.* **2019**, *44*, 3410–3413. [CrossRef]

21. He, K.; Zhang, X.; Ren, S.; Sun, J. Deep Residual Learning for Image Recognition. In Proceedings of the 2016 IEEE Conference on Computer Vision and Pattern Recognition (CVPR), Seattle, WA, USA, 27–30 June 2016; pp. 770–778. [CrossRef]

22. Baumeister, T.; Brunton, S.L.; Kutz, J.N. Deep learning and model predictive control for self-tuning mode-locked lasers. *J. Opt. Soc. Am. B Opt. Phys.* **2018**, *35*, 617–626. [CrossRef]

23. An, Y.; Huang, L.; Li, J.; Leng, J.; Yang, L.; Zhou, P. Learning to decompose the modes in few-mode fibers with deep convolutional neural network. *Opt. Express* **2019**, *27*, 10127–10137. [CrossRef] [PubMed]

24. Liu, Z.; Yan, S.; Liu, H.; Chen, X. Superhigh-Resolution Recognition of Optical Vortex Modes Assisted by a Deep-Learning Method. *Phys. Rev. Lett.* **2019**, *123*, 183902. [CrossRef] [PubMed]

25. Zibar, D.; Piels, M.; Jones, R.; Schaeeffer, C.G. Machine Learning Techniques in Optical Communication. *J. Lightwave Technol.* **2016**, *34*, 1442–1452. [CrossRef]

26. Genty, G.; Salmela, L.; Dudley, J.M.; Brunner, D.; Kokhanovskiy, A.; Kobtsev, S.; Turitsyn, S.K. Machine learning and applications in ultrafast photonics. *Nat. Photonics* **2020**, *15*, 91–101. [CrossRef]

27. Russakovsky, O.; Deng, J.; Su, H.; Krause, J.; Satheesh, S.; Ma, S.; Huang, Z.; Karpathy, A.; Khosla, A.; Bernstein, M.; et al. ImageNet Large Scale Visual Recognition Challenge. *Int. J.Comput. Vis.* **2015**, *115*, 211–252. [CrossRef]

28. Huang, G.; Liu, Z.; Van der Maaten, L.; Weinberger, K.Q. Densely Connected Convolutional Networks. *arXiv* **2017**, arXiv:1608.06993.

29. Yu, D.; Seltzer, M.L.; Li, J.; Huang, J.T.; Seide, F. Feature Learning in Deep Neural Networks—Studies on Speech Recognition Tasks. *arXiv* **2013**, arXiv:1301.3605.

30. Huang, G.; Liu, Z.; Pleiss, G.; van der Maaten, L.; Weinberger, K.Q. Convolutional Networks with Dense Connectivity. *arXiv* **2020**, arXiv:2001.02394.

31. Lee, C.Y.; Xie, S.; Gallagher, P.W.; Zhang, Z.; Tu, Z. Deeply-Supervised Nets. In Proceedings of the Artificial Intelligence and Statistics (AISTATS), San Diego, CA, USA, 9–12 May 2015; Volume 38, pp. 562–570.

32. Wang, Z.; Wang, Z.; Liu, Y.; He, R.; Wang, G.; Yang, G.; Han, S. Generation and time jitter of the loose soliton bunch in a passively mode-locked fiber laser. *Chin. Opt. Lett.* **2017**, *15*, 080605. [CrossRef]

33. He, K.; Zhang, X.; Ren, S.; Sun, J. Identity Mappings in Deep Residual Networks. In Proceedings of the 14th European Conference on Computer Vision (ECCV), Amsterdam, The Netherlands, 8–16 October 2016; Leibe, B., Matas, J., Sebe, N., Welling, M., Ed.; Lecture Notes in Computer Science; Springer: Cham, Switzerland, 2016; Volume 9908, pp. 630–645. [CrossRef]

34. Machrisaa, C. tensorflow-vgg: VGG19 and VGG16 on Tensorflow. Available online: https://github.com/machrisaa/tensorflow-vgg (accessed on 10 February 2021).

35. Simonyan, K.; Zisserman, A. Very deep convolutional networks for large-scale image recognition. *arXiv* **2014**, arXiv:1409.1556.

36. Li, Y.; Yuan, Y. Convergence Analysis of Two-layer Neural Networks with ReLU Activation. *Adv. Neural Inf. Process. Syst.* **2017**, *30*, 597–607.

37. Ioffe, S.; Szegedy, C. Batch Normalization: Accelerating Deep Network Training by Reducing Internal Covariate Shift. *arXiv* **2015**, arXiv:1502.03167.
38. Kingma, D.P.; Ba, J. Adam: A method for stochastic optimization. *arXiv* **2014**, arXiv:1412.6980.
39. Abadi, M.; Barham, P.; Chen, J.; Chen, Z.; Davis, A.; Dean, J.; Devin, M.; Ghemawat, S.; Irving, G.; Isard, M.; et al. TensorFlow: A system for large-scale machine learning. In Proceedings of the OSDI'16: 12th USENIX Symposium on Operating Systems Design and Implementation (OSDI), Savannah, GA, USA, 2–4 November 2016; pp. 265–283.
40. Meng, F.; Lapre, C.; Billet, C.; Genty, G.; Dudley, J.M. Instabilities in a dissipative soliton-similariton laser using a scalar iterative map. *Opt. Lett.* **2020**, *45*, 1232–1235. [CrossRef]

Article

Optical Machine Learning Using Time-Lens Deep Neural NetWorks

Luhe Zhang [1], **Caiyun Li** [1], **Jiangyong He** [1], **Yange Liu** [1], **Jian Zhao** [2], **Huiyi Guo** [1], **Longfei Zhu** [1], **Mengjie Zhou** [1], **Kaiyan Zhu** [1], **Congcong Liu** [1] and **Zhi Wang** [1,*]

[1]   Tianjin Key Laboratory of Optoelectronic Sensor and Sensing Network Technology, Institute of Modern Optics, Nankai University, Tianjin 300350, China; 2120180269@mail.nankai.edu.cn (L.Z.); 1120180099@mail.nankai.edu.cn (C.L.); 1120190097@mail.nankai.edu.cn (J.H.); ygliu@nankai.edu.cn (Y.L.); guohuiyi@mail.nankai.edu.cn (H.G.); 2120180259@mail.nankai.edu.cn (L.Z.); 2120190246@mail.nankai.edu.cn (M.Z.); 2120190247@mail.nankai.edu.cn (K.Z.); 1120200110@mail.nankai.edu.cn (C.L.)

[2]   Key Laboratory of Opto-Electronic Information Technical Science of Ministry of Education, School of Precision Instruments and Optoelectronics Engineering, Tianjin University, Tianjin 300372, China; enzhaojian@tju.edu.cn

*   Correspondence: zhiwang@nankai.edu.cn; Tel.: +86-139-2009-1078

Abstract: As a high-throughput data analysis technique, photon time stretching (PTS) is widely used in the monitoring of rare events such as cancer cells, rough waves, and the study of electronic and optical transient dynamics. The PTS technology relies on high-speed data collection, and the large amount of data generated poses a challenge to data storage and real-time processing. Therefore, how to use compatible optical methods to filter and process data in advance is particularly important. The time-lens proposed, based on the duality of time and space as an important data processing method derived from PTS, achieves imaging of time signals by controlling the phase information of the timing signals. In this paper, an optical neural network based on the time-lens (TL-ONN) is proposed, which applies the time-lens to the layer algorithm of the neural network to realize the forward transmission of one-dimensional data. The recognition function of this optical neural network for speech information is verified by simulation, and the test recognition accuracy reaches 95.35%. This architecture can be applied to feature extraction and classification, and is expected to be a breakthrough in detecting rare events such as cancer cell identification and screening.

Keywords: optical neural networks; time lens; fiber; dispersion Fourier transform; high-flux imaging; classification; cancer cell recognition; photon time stretching (PTS)

Citation: Zhang, L.; Li, C.; He, J.; Liu, Y.; Zhao, J.; Guo, H.; Zhu, L.; Zhou, M.; Zhu, K.; Liu, C.; et al. Optical Machine Learning Using Time-Lens Deep Neural NetWorks. *Photonics* **2021**, *8*, 78. https://doi.org/10.3390/photonics8030078

Received: 25 February 2021
Accepted: 12 March 2021
Published: 15 March 2021

1. Introduction

Recently, artificial neural networks (ANNs) have achieved significant developments rapidly and extensively. As the fastest developing computing method of artificial intelligence, deep learning has made remarkable achievements in machine vision [1], image classification [2], game theory [3], speech recognition [4], natural language processing [5], and other aspects. The use of elementary particles for data transmission and processing can lead to smaller equipment, greater speed, and lower energy consumption. The electron is the most widely used particle to date, and has become the cornerstone of the information society in signal transmission (cable) and data processing (electronic computer). Artificial intelligence chips represented by graphics processing units (GPUs), application-specific integrated circuits (ASICs), and field programmable gate arrays (FPGAs) have enabled electronic neural networks (ENNs) to achieve high precision, high convergence regression, and predict task performance [6]. When dealing with tasks with high complexity and high data volume, insurmountable shortcomings have emerged in ENNs, such as long time delay and low power efficiency caused by the interaction of many parameters in the network with the storage modules of electronic devices.

Fortunately, as a kind of boson, the photon has faster speed and lower energy consumption, resulting in it being significantly better than electrons in signal transmission and processing, and it has become a strong competitor for the elementary particles used in the next generation of information technology. Development of all-optical components, photonic chips, interconnects, and processors will bring the speed of light, photon coherence properties, field confinement and enhancement, information-carrying capacity, and the broad spectrum of light into the high-performance computing, the internet of things, and industries related to cloud, fog, and recently edge computing [7]. Due to the parallel characteristics of light in propagation, light interference, diffraction, and dispersion, phenomena can easily simulate various matrix linear operations, which are similar to the layer algorithm of forward propagation in neural networks. To pursue faster operating speed and higher power efficiency in information processing, the optical neural network (ONN), which uses photons as the information carrier, came at the right moment. Various ONN architectures have been proposed, including the optical interference neural network [8], the diffractive optical neural network [9–12], photonic reservoir computing [13,14], the photonic spiking neural network [15], and the recurrent neural network [16]. To process high-throughput and high-complexity data in real time, the algorithms in ONNs must have the characteristics of real-time information collection and rapid information measurement.

Photon time stretching (PTS), also known as dispersive Fourier transform technology (DFT), is a high-throughput real-time information collection technology that has emerged in recent years [17]. PTS can overcome the limitations of electronic equipment bandwidth and sampling speed, thus being able to realize ultra-fast information measurement, and its imaging frame rate is mainly determined by the mode-locked laser, which can reach tens of MHz/s or even GHz/s. DFT is widely used in ultra-high-speed microscopic imaging, microwave information analysis, spectral analysis, and observation of transient physical processes such as dissipative soliton structure, relativistic electron clusters, and rough waves [18]. It is worth emphasizing that this architecture plays an important role in the capture of rare events such as the early screening of cancer cells with large data volume characteristics. DFT broadens the pulse-carrying cell characteristics in the time domain and maps spectral information to the time domain; then, the information of the stretched light pulse is obtained through photo detection and a high-speed analog-to-digital converter, and finally the information is input into a computer or a special data signal processing chip for data processing. In 2009, researchers in the United States first proposed a method to achieve ultrafast imaging using PTS technology [19]. They then combined ultra-fast imaging and deep learning technology to distinguish colon cancer cells in the blood in 2016 [20]. In 2017, researchers from the University of Hong Kong reduced the monitoring of phytoplankton communities and used support vector machines to classify them, which can detect 100,000–1,000,000 cells per second [21]. In biomedicine, the combination of DFT and optical fluidics technology can complete high-flux imaging of hundreds of thousands to millions of cells per second, including various conditions in human blood and algae cells. It has great significance in cell classification [20], early cell screening [22–24] and feature extraction [25–29].

The high-throughput characteristics of PTS technology will inevitably produce a lot of data. Typically, the amount of data generated by a PTS system can reach 1 Tbit per second, which brings huge challenges to data storage and processing based on electronic devices and limits the application scope of this technology [30]. The high-throughput data generation of the DFT and the high-throughput processing characteristics of the photon neural network are perfectly complementary. Based on this characteristic, we propose a new architecture combining time-lens with the optical neural network (TL-ONN). According to the optical space–time duality [31] (that is, the spatial evolution characteristics of the light field caused by the diffraction effect and the time evolution characteristics of the optical pulse caused by the dispersion effect are equivalent), the imaging of the time signal can be realized by controlling the phase information of the timing signal, namely the time-lens. We establish a numerical model for simulation analysis to verify the feasibility of this architecture. By training 20,000 sets of speech data, we obtained a stable 98% recognition accuracy within one training cycle, which has obvious advantages of faster convergence and stable recognition accuracy compared with a deep neural network (DNN) with the same number of layers. This architecture implemented with all-optical components will offer outstanding improvements in biomedical science, cell dynamics, nonlinear optics, green energy, and other fields.

Here, we first introduce the architectural composition of the proposed TL-ONN, and then combine the time-lens principle with the neural network to drive the forward propagation and reverse optimization process. Finally, we use a speech dataset to train the proposed TL-ONN, and use numerical calculation to verify the classification function of this architecture.

2. Materials and Methods

The proposed ONN combines the conventional neural network with time stretch, realizing the deep learning function based on optics. As shown in Figure 1, two kinds of operations—time-lens transform and matrix multiplication—must be performed in each layer. The core optical structure which adapts the time-lens method is used to implement the first linear computation process. After that, the results are modulated by a weights matrix. Finally, the outputs serve as the input vector in the next layer. After calculation by the neural network composed of multiple time-lens layers, all input data are probed by a detector in the output layer. The prediction data and the target output are calculated by the cost function, and the gradient descent algorithm is carried out for each weights matrix (W_2) from backward propagation to achieve the optimal neural network structure. The input data of this network structure are generally one-dimensional time data. In the input layer, the physical information at each point in the time series is transferred to the neurons in each layer. Through the optical algorithm, each neuron between the layers is transmitted to realize the information processing behavior of the neural network.

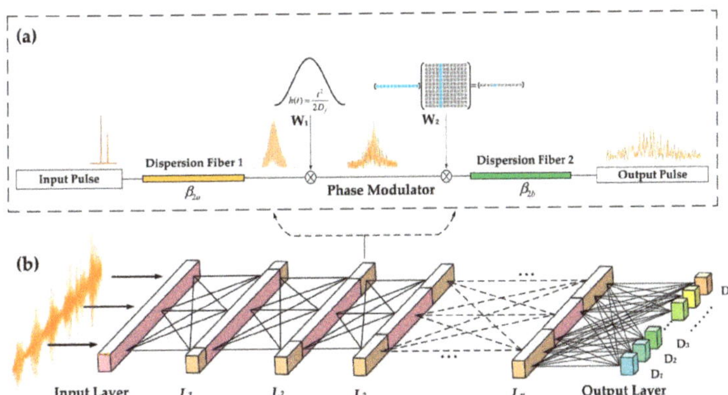

Figure 1. Optical neural network structure based on time-lens (TL-ONN). (**a**) The time-lens layer algorithm. The input data first pass through the first part of the dispersion fiber, undergoing phase modulation W_1, W_2 after the dispersion Fourier transform; the modulator reaches the optimal solution of the network after deep learning, and finally passes through the second segment of the dispersion fiber to complete the data transmission of each time-lens layer. β_{2a}, β_{2b}—the group-velocity dispersion of fiber 1 and fiber 2, respectively. W_1, W_2—the phase modulations. (**b**) TL-ONN structure. It comprises multiple time-lens layers. All time points on one layer can be regarded as neurons, and the neurons are transmitted through dispersion. L_1, L_2, ... , L_n—layers. D_1, D_2, ... , D_n—detectors.

Like the diffraction of space light, the time-lens plays a role of dispersion in time. As a result, the time-lens [32] can realize the imaging of the light pulse on the time scale. This is similar to the idea that the neurons in each layer of the neural network are derived from each neuron in the previous layer through a specific algorithm. The amplitude and phase of each point of the pulse after the time-lens is derived from the previous pulse calculated for each point. Based on this algorithm, an optical neural network based on the time lens is designed. Each neural network layer is formed by two segments of dispersive fiber and a second-order phase factor. The two layers are transmitted through intensity or a phase modulator. After backward propagation, each modulation factor is optimized by the gradient descent algorithm to obtain the best architecture.

2.1. Time-Lens Principle and Simulation Results

Analogous to the process by which a thin lens can image an object in space, a time-lens can image sequences in the time domain, such as laser pulses and sound sequences. In this section, we will introduce the principle of a time-lens starting from the propagation of narrow-band light pulses.

Assuming that the propagation area is infinite, the electric field envelop $\vec{E}(x,y,z,t)$ of a narrow-band laser pulse with a center frequency of ω_0 propagation in space coordinates (x,y,z) and time t satisfies

$$\vec{E}(x,y,z,t) = \vec{A}(x,y)e^{i(\omega_0 t + \beta(\omega_0)z)} \tag{1}$$

where $\vec{A}(x,y)$ is the electric field envelope of the input light pulse, $\beta(\omega_0)$ is the dispersion coefficient, and ω represents the angular frequency. Expanding the dispersion coefficient $\beta(\omega)$ with Taylor series and retaining it to the second order, the frequency spectrum $\Lambda(z,\omega)$ after Fourier transformation can be described as

$$\frac{\partial \Lambda(z,\omega-\omega_0)}{\partial z} = -i\left[(\omega-\omega_0)\frac{d\beta}{d\omega} + \frac{(\omega-\omega_0)^2}{2}\frac{d^2\beta}{d\omega^2}\right]\Lambda(z,\omega-\omega_0), \tag{2}$$

Then, we perform the inverse Fourier transform on (2) to obtain the time domain pulse envelope:

$$\frac{\partial A(z,t)}{\partial z} + \frac{1}{V_g}\frac{\partial A(z,t)}{\partial t} = \frac{i}{2}\frac{d^2\beta}{d\omega^2}\frac{\partial^2 A(z,t)}{\partial t^2}, \tag{3}$$

where V_g is the group velocity, $V_g = \frac{d\omega}{d\beta}$. If we establish a new coordinate whose frame moves at the speed of the group velocity of light, the corresponding transformation can be described as

$$T = (t-t_0) - \frac{z-z_0}{V_g}, \tag{4}$$

$$Z = z - z_0, \tag{5}$$

where t_0 and z_0 are the time and the space initial points, respectively. Under this circumstance, (3) can be simplified as

$$\frac{\partial A(Z,T)}{\partial Z} = \frac{1}{2}\frac{d^2\beta}{d\omega^2}\frac{\partial^2 A(Z,T)}{\partial T^2}, \tag{6}$$

Then, we can get the spectrum of the signal envelope by Fourier transform:

$$\Lambda(Z,\tau) = \Lambda(0,\tau)\exp\left(-\frac{iZ\beta''}{2}\omega^2\right). \tag{7}$$

where τ is the time variable in frequency domain, i is the imaginary number. It can be seen from the time domain envelope equation that the second order phase modulation of the independent variable T is carried out in the time-lens algorithm. Like the space lens, the space diffraction equation of a paraxial beam and the propagation equation of a narrow-band optical pulse in the dispersive medium both modulate the independent variable (x, y, and t) second order.

The time lens mainly comprises three parts—the second-order phase modulator and the dispersion medium before and after the modulator (Figure 2a). In the dispersion medium part, the pulse passing through the long distance dispersion fiber is equivalent to the pulse being modulated in the frequency domain by a factor determined by the fiber length and the second-order dispersion coefficient, which can be expressed as $G_i(Z_i, \omega) = \exp\left(-i\frac{Z_i}{2}\beta_{2i}\omega^2\right)$ where Z_i and β_{2i} represent the length of fiber i and the second-order dispersion coefficient, respectively. When passing through the time domain phase modulator, the phase factor satisfying the imaging condition of the time-lens is the quadratic function of time τ by $\varphi_{\text{timelens}}(\tau) = \exp\left(i\frac{\tau^2}{2D_f}\right)$, and D_f is the focal length of the time-lens satisfying the imaging conditions of the time-lens. With respect to analog space-lens imaging conditions, the time-lens imaging condition is

$$\frac{1}{Z_1\beta_{2a}/2} + \frac{1}{Z_2\beta_{2b}/2} = -\frac{1}{D_f/2\omega_0},\tag{8}$$

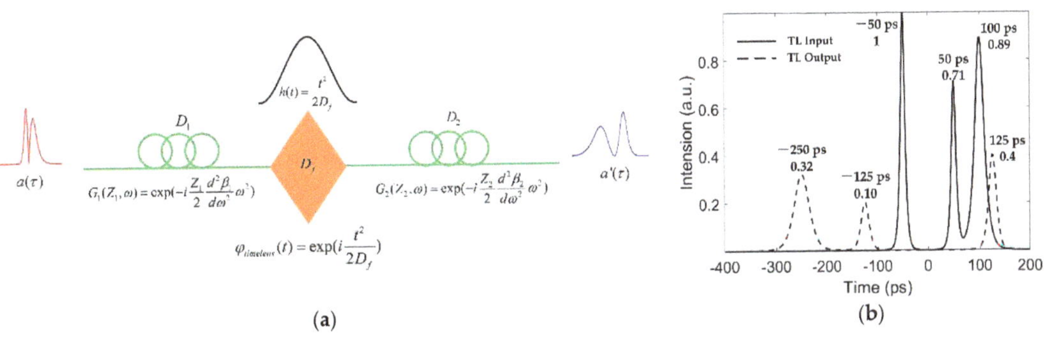

(a)

(b)

Figure 2. Time-lens principle and imaging of soliton pulses. (a) The imaging of the pulse by the time lens mainly comprises two dispersive fibers and the secondary phase modulation factors with respect to time t. $a(\tau)$ and $a'(\tau)$ represent the pulse envelope before and after transmission through the time-lens, respectively. D_1 and D_2 represent two dispersion fibers. G_1 and G_2 are the transmission function of dispersive fibers 1 and 2 in the frequency domain, respectively. $h(t)$ is a function with respect to the square of time, and constitutes the second order phase modulat or $\varphi_{\text{timelens}}(t)$ in the time-lens. (b) Example of time-lens imaging of pulse. The peak position (top) and normalized intensity (bottom) of each pulse are identified in the figure.

Its magnification can be expressed as $M = -Z_1\beta_{2b}/Z_2\beta_{2a}$ (see Appendix A). Figure 2b shows a comparison of the duration of a group of soliton pulses and their output of the time lens at M = 2.5; the peak position and normalized intensity of the pulse are marked to verify its magnification. In summary, after passing through the time-lens, the pulse is $\frac{1}{\sqrt{M}}$ times larger in amplitude and M times larger in duration, and a second order phase modulation is added in phases.

2.2. Mathematical Analysis of TL-ONN

In this section, we will analyze the transmission process of input data in two adjacent time-lens layers. Suppose that the input pulse can be expressed as A(0, t), that is, the initial intensity in time of the pulse into the first dispersion fiber of the time lens. The intensity of the input data at each time point will be mapped to all time points according to a specific algorithm after two segments of the dispersion fiber in the time lens and second-order phase modulation in the time domain. Equation (9) shows the algorithm results; its derivation can be found in Appendix A. In the neural network based on this algorithm, each neuron in the *mth* layer can be regarded as the result of mapping all neurons in the *(m − 1)th* layer.

$$A_{t_i}^l = \frac{1}{2\pi\sqrt{M}} \exp\left(j \frac{\omega_0 \left(t_k^{l-1} \right)^2}{2MD_f} \right) \int_{-\infty}^{+\infty} \Lambda(0,\omega') \exp\left(-i \frac{t_k^{l-1}}{M} \omega' \right) d\omega', \tag{9}$$

where $M = -\frac{Z_1}{2}\frac{d^2\beta_b}{d\omega^2} / \frac{Z_2}{2}\frac{d^2\beta_f}{d\omega^2}$ represents the magnification factor of the time-lens, β_b and β_f are the second-order dispersion coefficients of the two segments of the dispersion fiber, Z_1 and Z_2 are the lengths of the two segments of the dispersion fiber, l represents the layer number, t_k represents all neurons that contribute to the neuron t_i in the *lth* layer.

The intensity and phase of the neuron t_i in the L layer are determined by both the input pulse in the L − 1 layer and the modulation coefficient in the L layer. For the Lth layer of the network, the information on each neuron can be expressed by

$$n_{t_i}^l = h_{t_i}^l \cdot \sum_k n_{k,t_i}^{l-1}, \tag{10}$$

where $m_{t_i}^l = \sum_k n_{k,t_i}^{l-1}$ is the input pulse to neuron t_i of layer l, n_{k,t_i}^{l-1} represents the contribution of the k-th neuron of the layer $l − 1$ to the neuron t_i of the layer l. $h_{t_i}^l$ is the modulation coefficient of the neuron t_i in layer l; the modulation coefficient of a neuron comprises amplitude and phase items, i.e., $h_{t_i}^l = a_{t_i}^l \exp\left(j\phi_{t_i}^l \right)$.

The forward model of our TL-ONN architecture is illustrated in Figure 1 and notated as follows:

$$\begin{cases} n_{t_i}^l = h_{t_i}^l \cdot m_{t_i}^l \\ m_{t_i}^l = \sum_k n_{k,t_i}^{l-1} \\ h_{t_i}^l = a_{t_i}^l exp\left(j\phi_{t_i}^l \right) \end{cases}, \tag{11}$$

where t_i refers to a neuron of the *l*th layer, and k refers to a neuron of the previous layer, connected to neuron t_i by optical dispersion. The input pulse n_k^0, which is located at layer 0 (i.e., the input plane), is in general a complex-valued quantity and can carry information in its phase and/or amplitude channels.

Assuming that the TL-ONN design is composed of N layers (excluding the input and output planes), the data transmitted through the architecture are finally detected by PD, and detectors are placed at the output plane to measure the intensity of the output data. If the bandwidth of the PD is much narrower than the output signal bandwidth, the PD will serve not only as an energy transforming device but also as a pulse energy accumulator. The final output of the architecture can be expressed as

$$s^{N+1} = \sum_{t_i} w_{ti} \cdot n_{t_i}^N, \tag{12}$$

where $n_{t_i}^N$ represents the neuron t_i of the output layer (N), and w_{t_i} is the energy accumulation coefficient of PD on the time axis of the data.

To train a TL-ONN design, we used the error back-propagation algorithm along with the stochastic gradient descent optimization method. A loss function was defined to evaluate the performance of the network parameters to minimize the loss function. Without loss of generality, here we focus on our classified architecture and define the loss function (E) using the cross-entropy error between the output plane intensity s^{N+1} and the target g^{N+1}:

$$E\left(a_{t_i}^l, \phi_{t_i}^l \right) = \frac{1}{k} \sum_k \left[g^{N+1} ln s^{N+1} + \left(1 - g^{N+1} \right) \ln\left(1 - s^{N+1} \right) \right] \tag{13}$$

In the network based on a time-lens algorithm consisting of N time-lens layers, the data characteristics in the previous layer with α neurons are extracted into neurons in the current layer with β neurons, where $\beta = \alpha \cdot k_{L-1,L}$ and $k_{L-1,L}$ represents the scaling multiples between the $(L - 1)$th layer and the Lth layer. The time-lens algorithm has a similar function of removing the redundant information and compressing the features as the pooling layer in a conventional ANN. The characteristics carried by the input data will emerge and be highlighted through each layer after being transmitted through this classification architecture, and finally evolve into the labels of the corresponding category.

3. Results

In order to verify the effectiveness of the system in the time-domain information classification, we used numerical methods to simulate the TL-ONN to realize the recognition of specific sound signals. We used a dataset containing 18,000 training data and 2000 test data picked from intelligent speech database [33] to evaluate the performances of TL-ONN. The content in the speech dataset is the wake phrase "Hi, Miya!" in English and Chinese collected in the actual home environment using a microphone array and Hi-Fi microphone. The test subset provides paired target/non-target answers to evaluate verification results. In general, we used the dichotomy problem to test the classification performances of two kinds of systems including the TL-ONN and the conventional DNN.

We first constructed a TL-ONN composed of five time-lens layers to verify the classification feasibility of this architecture. Figure 3a shows the training results of TL-ONN in the cases of $k_{L-1,L} = 0.6$. The accuracy of the TL-ONN for a total of 2000 test samples is above 98% (Figure 3a top), which is close to the accuracy for the DNN (Figure 3a bottom). The horizontal axis represents the number of training steps in one training batch (batch size = 50). The accuracy of this test fluctuates greatly in the first few steps, and then reaches over 98% at about 17 steps and remains stable. In contrast, it was difficult for a five-layer DNN network under the same conditions to achieve stable accuracy and training loss in one epoch (Figure 3a). When the training epoch was set to 10, it was found that the test accuracy and training loss still changed suddenly at the 10th training epoch, which might be due to gradient explosion, overfitting, or another reason. We define the accuracy as the proportion of the number of output labels that are the same as the target label to the total number of test sets. Using the same 2000 test set to test the two networks' architecture, the accuracy rates reached 95.35% (Figure 3b) and 93.2% (Figure 3c). In general, TL-ONN has significant advantages over DNN in verifying classification performance.

To easily see the changes of the two types of voice information in each layer of TL-ONN, we extracted two sets of input with typical characteristics for observation. Figure 4a shows the layer structure of this network, which contains multiple time-lens layers, where each time point on a given layer acts as a neuron with a complex dispersion coefficient. Figure 4b,c shows the data evolution of each layer when two types of speech are input to the network. From the input layer, we can distinguish the differences between the two types of input data from the shape of the waveform. The waveform containing "Hi, Miya!" has a higher continuity, while the waveform of random speech has quantized characteristics and always has a value on the time axis. On the second layer of the network, the "Hi, Miya!" input will change into several sets of pulses through the time-lens layer and another type of information will spread all over the time. After being transmitted by multiple time-lens layers, the two inputs will eventually change to the shape in Layer 6, and the two types of speech will eventually evolve into the shape of the impact function at different time points. As shown in Figure 4b,c, D1 and D2 correspond to detectors of different input types. The random speech eventually responds at D1, while the input containing "Hi, Miya!" responds at D2.

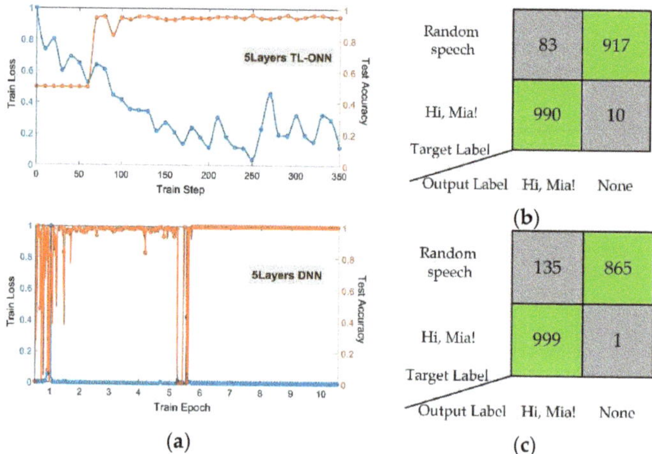

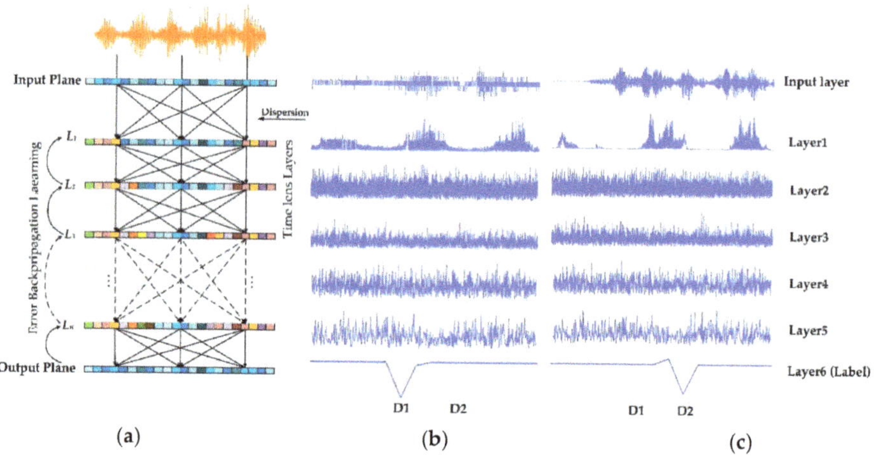

Figure 3. (**a**) Change curves of the loss function and accuracy of training TL-ONN (top) and deep neural network (DNN) (bottom) with 18,000 sets of speech data each. (**b,c**) Statistical results of the number of correct (green squares) and incorrect (grey squares) output label after the training of the two networks' architecture is completed. We define the accuracy rate as the percentage of the correct result in the total test set data (2000).

Figure 4. TL-ONN layer structure and the change process of input information at each layer. (**a**) The layer structure of this network, which contains multiple time-lens layers, where each time point on a given layer acts as a neuron with a complex dispersion coefficient. L_1, L_2,..., L_n represent the time lens layer, and D1 and D2 represent two types of detectors on the output plane. Different colors are used to distinguish neurons that carry different messages. The evolution of the two types of input data in each layer of the structure ((**b,c**) contain "Hi, Miya!" and random speech, respectively), and the two inputs are responses at different detectors in the output layer.

To eliminate the contingency of the experiment, we set up a series of networks consisting of 3–8 layers to test the influence of different numbers of time-lens layers on classification performance. Figure 5 shows the test results of the TL-ONN architecture composed of different numbers of time-lens layers—33, 30, and17 steps are needed in the TL-ONN with three, four, and five layers, respectively, to reach an accuracy of 98% (Figure 5a). When the number of time-lens layers is increased to six or more, the accuracy can be stabilized at 98–99% after about 10 training steps; however, an unlimited increase in the number of time-lens layers does not make the results of network training infinitely better. For example, we can see that compared with a network with six, seven, or eight layers, TL-ONN requires more steps to achieve stable accuracy. Overall, the network with six time-lens layers has the best classification performance. All the results discussed above occur in one training epoch. At least a few epochs were needed to achieve stable classification accuracy for conventional DNN with the same dataset. TL-ONN has obvious advantages of faster convergence speed and stable classification accuracy.

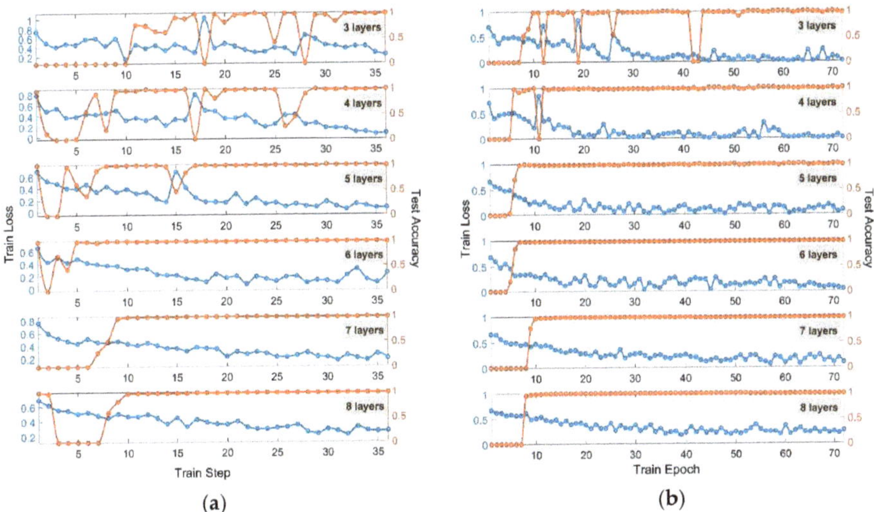

Figure 5. Optical neural network classification performance based on the time-lens. The orange line represents training accuracy, and the blue one is training loss. (**a**) Training result of a series of TL-ONN composed of 3–8 time lens layers. (**b**) Training result of TL-ONN after exchange of two phase modulators consisting of 3–8 time lens layers.

Similarly, we reverse the order of the phase modulator W_1 and W_2, and use the same training set for training. Figure 5b shows the classification results under this architecture, and the time-scaling multiple between two layers is still 0.6. Under the same conditions, a series of networks consisting of three–eight layers were constructed to test the classification performance. To achieve an accuracy of 98%, 55 and 12 steps are needed in the TL-ONN with three and four layers, respectively. The accuracy can be stabilized at 98–99% after about 10 training steps when the number of time-lens layers is increased to five or more. As with the previous results, compared with a network with six, seven, or eight layers, TL-ONN requires more steps to achieve stable accuracy. Overall, the network structure with six time-lens layers has the best classification performance, and it is consistent with the results of the former architecture.

At the detector/output plane, we measured the intensity of the network output, and as a loss function to train the classification TL-ONN, we used its mean square error (MSE) against the target output. The classification of TL-ONN was trained using a modulator (W_2), where we aimed to maximize the normalized signal of each target's corresponding detector region, while minimizing the total signal outside of all the detector regions. We used the stochastic gradient descent algorithm, Adam [34], to back-propagate the errors and update the layers of the network to minimize the loss function. The classifier TL-ONN was trained with speech datasets [33], and achieved the desired mapping functions between the input and output planes after five steps. The training batch size was set to be 50 for the speech classifier network. To verify the feasibility of the TL-ONN architecture, we used the python language to establish a simulation model for theoretical analysis. The networks were implemented using Python version 3.8.0. and PyTorch version 1.4.0. Using a desktop computer (GeForce GTX 1060 Graphical Processing Unit, GPU and Intel(R) Core (TM) i7-8700 CPU @3.20GHz and 64GB of RAM, running a Windows 10 operating system, Microsoft), the above-outlined PyTorch-based design of a TL-ONN architecture took approximately 26 h to train for the classifier networks.

Compared with conventional DNNs, TL-ONN is not only a physical and optical neural network but also has some unique architecture. First, the time-lens algorithm applied at each layer of the network can refine the features of the input data, similar to what is used as a pooling layer, remove redundant information, and compress features. The time-lens method can be regarded as the pooling element in the photon. Second, TL-ONN can handle complex values, such as complex nonlinear dynamics in passively mode-locked lasers. The phase modulators can respectively modulate different physical parameters, and as long as the modulator parameters are determined, a passive all-optical neural network can be basically realized. Third, the output of each neuron is coupled to the neurons in the next layer through a certain weight relationship through the dispersion effect of the optical fiber, thereby providing a unique interconnection from within the network.

4. Discussion

In this paper, we proposed a new optical neural network based on the time-lens method. The forward transmission of the neural network can be realized by the time lens to enlarge or reduce the data in the time dimension, and the characteristics of the signal extracted by the time-lens algorithm are modulated with the amplitude or phase modulator to realize the weight matrix optimization process in linear operation. After the time signal is compressed and modulated by the multilayer based on the time-lens method, it will eventually evolve into the corresponding target output, so as to realize the classification function of the optical neural network. To verify the feasibility of the network, we used the speech data set to train it and got a test accuracy of 95.35%. The accuracy is obviously more stable and has faster convergence compared with the same number of layers in a DNN.

Our optical architecture implements a feedforward neural network through a time-stretching method; thus, when completing high-throughput data processing and large-scale tasks, it basically proceeds at the speed of light in the optical fiber, and requires little additional power consumption. The system has a clear correspondence between the theoretical neural network and the actual optical component parameters; thus, once each parameter in the network can be optimized, it can basically be realized completely by optical devices, which provides the possibility of building an all-optical neural network test system composed of optical fibers, electro-optic modulators, etc.

Here, we verify the feasibility of the proposed TL-ONN by numerical simulation, and we will work to build a test system to realize all-optical TL-ONN in the future. It is often accompanied by noise and loss in experiments. We conservatively speculate that such noise may reduce the classification accuracy of the architecture. On the other hand, in order to solve the influence of loss on the experiment, an optical amplifier is generally added to improve the signal-to-noise ratio. The non-linear effects of the optical amplifier have similar functions to the activation function in the neural network, and it may play an important role in all-optical neural networks in the future.

The emergence of ONNs provides a solution for real-time online processing of high-throughput timing information. By fusing the ONN with the photon time stretching test system, not only can real-time data processing be achieved, but also the system's dependence on broadband high-speed electronic systems can be significantly reduced. In addition, cost and power consumption can be reduced, and the system can be used in medicine and biology, green energy, physics, and optical communication information extraction, having more extensive applications. This architecture is expected to provide breakthroughs in the identification of rare events such as the initial screening of cancer cells and be widely used in high-throughput data processing such as early cell screening [22], drug development [23], cell dynamics [21], and environmental improvement [35,36], as well as in other fields.

Author Contributions: Conceptualization, L.Z. (Luhe Zhang) and Z.W.; methodology, L.Z. (Luhe Zhang), J.H. and Z.W.; software, L.Z. (Luhe Zhang), C.L. (Caiyun Li) and H.G.; validation, L.Z. (Luhe Zhang), J.H. and H.G.; formal analysis, L.Z. (Luhe Zhang), L.Z. (Longfei Zhu), Y.L., J.Z. and M.Z.; investigation, L.Z. (Luhe Zhang), C.L. (Congcong Liu) and K.Z.; resources, Z.W.; data curation, L.Z. (Luhe Zhang) and H.G.; writing—original draft preparation, L.Z. (Luhe Zhang); writing—review and editing, L.Z. (Luhe Zhang); visualization, L.Z. (Luhe Zhang) and H.G.; supervision, L.Z. (Luhe Zhang); project administration, Z.W.; funding acquisition, Z.W. All authors have read and agreed to the published version of the manuscript.

Funding: This research was funded by the National Key Research and Development Program of China under Grant No. 2018YFB0703500, the National Natural Science Foundation of China (NSFC) (Grant Nos. 61775107, 11674177, and 61640408), and the Tianjin Natural Science Foundation (Grant No. 19JCZDJC31200), China.

Institutional Review Board Statement: Ethical review and approval were waived for this study, due to this research only uses human voice as training data to verify the classification function of the TL-ONN we proposed, rather than studying the voice itself.

Informed Consent Statement: Informed consent was obtained from all subjects involved in the study.

Data Availability Statement: The data that support the plots within this paper and other findings of this study are available from the corresponding author upon reasonable request. The data processing and simulation codes that were used to generate the plots within this paper and other findings of this study are available from the corresponding author upon reasonable request.

Conflicts of Interest: The authors declare no conflict of interest.

Appendix A

In this section, we use the transmission function of each part of the time-lens to derive the output time domain envelope. Table A1 shows the transfer function of the two dispersion fibers in the frequency domain and the second phase modulation in the time domain. This is assuming that the time domain and frequency domain envelopes of the input pulse are $A(0, T)$ and $\Lambda(0, \omega)$, respectively, and the lengths of the two fibers are Z_1 and Z_2, respectively. After passing the fiber D1, the time domain of the pulse can be described as

$$A(Z_1, T) = f^{-1}\{\Lambda(0, \omega) \cdot G_1(Z_1, \omega)\}, \tag{A1}$$

after the second phase modulation with respect to time, the pulse becomes

$$A(Z_1 + \varepsilon, T) = f^{-1}\{\Lambda(0, \omega) \cdot G_1(Z_1, \omega)\} \cdot \varphi_{timelens}(t), \tag{A2}$$

Finally, the pulse passes through the fiber D2 and we can get the output of the time lens expressed in the time domain as

$$A(Z_1 + \varepsilon + Z_2, T) = \tfrac{1}{2\pi} f^{-1}\{[(\Lambda(0, \omega) \cdot G_1(Z_1, \omega)) * f\{\varphi_{timelens}(t)\}] \cdot G_2(Z_2, \omega)\}, \tag{A3}$$

where ε distinguishes the signal expression before and after the second-order phase modulation, f and f^{-1} represent the Fourier transform and the inverse Fourier transform, respectively. The time-lens output is mainly based on the time domain convolution theorem and frequency domain convolution theorem.

Table A1. Transmission function of the time-lens.

	Frequency Domain	Time Domain	Parameters
D_1	$G_1(Z_1, \omega) = \exp(-ia\omega^2)$	—	$a = \dfrac{Z_1}{2}\dfrac{d^2\beta_1}{d\omega^2}$
D_2	$G_2(Z_2, \omega) = \exp(-ib\omega^2)$	—	$b = \dfrac{Z_2}{2}\dfrac{d^2\beta_2}{d\omega^2}$
D_f	—	$\varphi_{timelens}(t) = \exp\left(i\dfrac{t^2}{4c}\right)$	$c = \dfrac{D_f}{2\omega_0}$

After inverse Fourier transformation of $\varphi_{timelens}(t)$, the frequency domain expression $\psi(\omega) = \sqrt{4\pi ic}\exp\left[-ic\omega^2\right]$ can be obtained. Using convolution calculation:

$$(\Lambda(0,\omega)\cdot G_1(Z_1,\omega)) * f\{\varphi_{timelens}(t)\} = \int_{-\infty}^{+\infty}\Lambda(0,\omega')\cdot G_1(Z_1,\omega')\cdot\psi(\omega-\omega')d\omega'. \quad (A4)$$

Putting (A4) into (A3) and switching the order of integration, the output of the time-lens can be written as

$$A(Z_1+\varepsilon+Z_2,T)$$
$$= \frac{1}{2\pi}\int_{-\infty}^{+\infty}\exp(i\omega T)G_2(Z_2,\omega)\psi(\omega-\omega')d\omega \quad (A5)$$
$$\cdot\frac{1}{2\pi}\int_{-\infty}^{+\infty}\Lambda(0,\omega')G_1(Z_1,\omega')d\omega',$$

substituting $G_2(Z_2,\omega)$ and $\psi(\omega-\omega')$ into the integral calculation of ω and performing the integral operation:

$$\frac{1}{2\pi}\int_{-\infty}^{+\infty}\exp(i\omega T)G_2(Z_2,\omega)\psi(\omega-\omega')d\omega = \sqrt{\frac{c}{b+c}}\exp\left(-ic\omega'^2\right)\exp\left[i\left(T+\frac{2c\omega'}{2\sqrt{b+c}}\right)^2\right] \quad (A6)$$

Bringing (A6) back to (A5), after merging similar items, the final output of the time-lens is described by

$$A(Z_1+\varepsilon+Z_2,T) =$$
$$\sqrt{\frac{c}{b+c}}\exp\left[i\left(\frac{T^2}{\left(2\sqrt{b+c}\right)^2}\right)\right]\cdot\frac{1}{2\pi}\int_{-\infty}^{+\infty}\Lambda(0,\omega')\exp\left[-i\left(a+c-\frac{c^2}{b+c}\right)\omega'^2\right]\exp\left[i\left(\frac{cT}{b+c}\omega'\right)\right]d\omega'. \quad (A7)$$

According to imaging conditions, if the time imaging system confirms that the variation is only found in size instead of shape between input and output pulse envelopes, it is necessary to confirm that the coefficient value of the quadratic term of ω' is equal to 1:

$$a+c-\frac{c^2}{b+c} = 0 \quad (A8)$$

Therefore, the integral term in (A7) can become an inverse Fourier transform, which is equivalent to $\Lambda\left(\frac{cT}{b+c},\omega'\right)$. Bring a,b,c into (A8) to get the imaging conditions of the time lens:

$$\frac{1}{\frac{Z_1}{2}\frac{d^2\beta_1}{d\omega^2}} + \frac{1}{\frac{Z_2}{2}\frac{d^2\beta_2}{d\omega^2}} = -\frac{1}{\frac{D_f}{2\omega_0}}, \quad (A9)$$

and the time magnification is defined by the first-order coefficient of ω':

$$M = \frac{c}{b+c} = -\frac{a}{b} = -\frac{\frac{Z_1}{2}\frac{d^2\beta_1}{d\omega^2}}{\frac{Z_2}{2}\frac{d^2\beta_2}{d\omega^2}}. \quad (A10)$$

Introducing the magnification factor into (A7), we can finally get the basis of the time lens layer algorithm (9):

$$A(Z_1+\varepsilon+Z_2,T) = \frac{1}{2\pi\sqrt{M}}exp\left(i\frac{\omega_0 T^2}{2MD_f}\right)\int_{-\infty}^{+\infty}\Lambda(0,\omega')\exp\left(-i\frac{T}{M}\omega'\right)d\omega' \quad (A11)$$

When the time-lens algorithm (A11) is applied to TL-ONN, each time point can be regarded as a neuron, and thus the calculation result of each neuron (9) in the *m*th layer is obtained.

References

1. Barbastathis, G.; Ozcan, A.; Situ, G. On the use of deep learning for computational imaging. *Optica* **2019**, *6*, 921. [CrossRef]
2. Wei, Y.; Xia, W.; Lin, M.; Huang, J.; Ni, B.; Dong, J.; Zhao, Y.; Yan, S. HCP: A flexible CNN framework for multi-label image classification. *IEEE Trans. Pattern Anal. Mach. Intell.* **2016**, *38*, 1901–1907. [CrossRef]
3. Silver, D.; Huang, A.; Maddison, C.J.; Guez, A.; Sifre, L.; Van Den Driessche, G.; Schrittwieser, J.; Antonoglou, I.; Panneershelvam, V.; Lanctot, M.; et al. Mastering the game of Go with deep neural networks and tree search. *Nature* **2016**, *529*, 484–489. [CrossRef]
4. Furui, S.; Deng, L.; Gales, M.; Ney, H.; Tokuda, K. Fundamental technologies in modern speech recognition. *IEEE Signal Process. Mag.* **2012**, *29*, 16–17. [CrossRef]
5. Abdi, A.; Shamsuddin, S.M.; Hasan, S.; Piran, J. Deep learning-based sentiment classification of evaluative text based on Multi-feature fusion. *Inf. Process. Manag.* **2019**, *56*, 1245–1259. [CrossRef]
6. Pei, J.; Deng, L.; Song, S.; Zhao, M.; Zhang, Y.; Wu, S.; Wang, G.; Zou, Z.; Wu, Z.; He, W.; et al. Towards artificial general intelligence with hybrid Tianjic chip architecture. *Nature* **2019**, *572*, 106–111. [CrossRef]
7. Passian, A.; Imam, N. Nanosystems, edge computing, and the next generation computing systems. *Sensors* **2019**, *19*, 48. [CrossRef]
8. Shen, Y.; Harris, N.C.; Skirlo, S.; Prabhu, M.; Baehr-Jones, T.; Hochberg, M.; Sun, X.; Zhao, S.; Larochelle, H.; Englund, D.; et al. Deep learning with coherent nanophotonic circuits. *Nat. Photonics* **2017**, *11*, 441–446. [CrossRef]
9. Lin, X.; Rivenson, Y.; Yardimci, N.T.; Veli, M.; Luo, Y.; Jarrahi, M.; Ozcan, A. All-optical machine learning using diffractive deep neural networks. *Science* **2018**, *361*, 1004–1008. [CrossRef]
10. Yan, T.; Wu, J.; Zhou, T.; Xie, H.; Xu, F.; Fan, J.; Fang, L.; Lin, X.; Dai, Q. Fourier-space Diffractive Deep Neural Network. *Phys. Rev. Lett.* **2019**, *123*, 23901. [CrossRef]
11. Chang, J.; Sitzmann, V.; Dun, X.; Heidrich, W.; Wetzstein, G. Hybrid optical-electronic convolutional neural networks with optimized diffractive optics for image classification. *Sci. Rep.* **2018**, *8*, 1–10. [CrossRef]
12. Luo, Y.; Mengu, D.; Yardimci, N.T.; Rivenson, Y.; Veli, M.; Jarrahi, M.; Ozcan, A. Design of task-specific optical systems using broadband diffractive neural networks. *Light Sci. Appl.* **2019**, *8*, 1–14. [CrossRef]
13. Antonik, P.; Marsal, N.; Rontani, D. Large-scale spatiotemporal photonic reservoir computer for image classification. *IEEE J. Sel. Top. Quantum Electron.* **2019**, *26*, 1–12. [CrossRef]
14. Larger, L.; Baylón-Fuentes, A.; Martinenghi, R.; Udaltsov, V.S.; Chembo, Y.K.; Jacquot, M. High-speed photonic reservoir computing using a time-delay-based architecture: Million words per second classification. *Phys. Rev. X* **2017**, *7*, 1–14. [CrossRef]
15. Hamerly, R.; Bernstein, L.; Sludds, A.; Soljačić, M.; Englund, D. Large-Scale Optical Neural Networks Based on Photoelectric Multiplication. *Phys. Rev. X* **2019**, *9*, 1–12. [CrossRef]
16. Hughes, T.W.; Williamson, I.A.D.; Minkov, M.; Fan, S. Wave physics as an analog recurrent neural network. *Sci. Adv.* **2019**, *5*, 1 7. [CrossRef]
17. Goda, K.; Jalali, B. Dispersive Fourier transformation for fast continuous single-shot measurements. *Nat. Photonics* **2013**, *7*, 102–112. [CrossRef]
18. Mahjoubfar, A.; Churkin, D.V.; Barland, S.; Broderick, N.; Turitsyn, S.K.; Jalali, B. Time stretch and its applications. *Nat. Photonics* **2017**, *11*, 341–351. [CrossRef]
19. Goda, K.; Tsia, K.K.; Jalali, B. Serial time-encoded amplified imaging for real-time observation of fast dynamic phenomena. *Nature* **2009**, *458*, 1145–1149. [CrossRef]
20. Chen, C.L.; Mahjoubfar, A.; Tai, L.C.; Blaby, I.K.; Huang, A.; Niazi, K.R.; Jalali, B. Deep Learning in Label-free Cell Classification. *Sci. Rep.* **2016**, *6*, 1–16. [CrossRef]
21. Tang, A.H.L.; Yeung, P.; Chan, G.C.F.; Chan, B.P.; Wong, K.K.Y.; Tsia, K.K. Time-stretch microscopy on a DVD for high-throughput imaging cell-based assay. *Biomed. Opt. Express* **2017**, *8*, 640. [CrossRef] [PubMed]
22. Guo, B.; Lei, C.; Kobayashi, H.; Ito, T.; Yalikun, Y.; Jiang, Y.; Tanaka, Y.; Ozeki, Y.; Goda, K. High-throughput, label-free, single-cell, microalgal lipid screening by machine-learning-equipped optofluidic time-stretch quantitative phase microscopy. *Cytom. Part A* **2017**, *91*, 494–502. [CrossRef] [PubMed]
23. Kobayashi, H.; Lei, C.; Wu, Y.; Mao, A.; Jiang, Y.; Guo, B.; Ozeki, Y.; Goda, K. Label-free detection of cellular drug responses by high-throughput bright-field imaging and machine learning. *Sci. Rep.* **2017**, *7*, 1–9. [CrossRef]
24. Lai, Q.T.K.; Lee, K.C.M.; Tang, A.H.L.; Wong, K.K.Y.; So, H.K.H.; Tsia, K.K. High-throughput time-stretch imaging flow cytometry for multi-class classification of phytoplankton. *Opt. Express* **2016**, *24*, 28170. [CrossRef]
25. Göröcs, Z.; Tamamitsu, M.; Bianco, V.; Wolf, P.; Roy, S.; Shindo, K.; Yanny, K.; Wu, Y.; Koydemir, H.C.; Rivenson, Y.; et al. A deep learning-enabled portable imaging flow cytometer for cost-effective, high-throughput, and label-free analysis of natural water samples. *Light Sci. Appl.* **2018**, *7*. [CrossRef] [PubMed]
26. Eulenberg, P.; Köhler, N.; Blasi, T.; Filby, A.; Carpenter, A.E.; Rees, P.; Theis, F.J.; Wolf, F.A. Reconstructing cell cycle and disease progression using deep learning. *Nat. Commun.* **2017**, *8*, 1–6. [CrossRef] [PubMed]
27. Mahmud, M.; Shamim Kaiser, M.; Hussain, A.; Vassanelli, S. Applications of deep learning and reinforcement learning to biological data. *IEEE Trans. Neural Netw. Learn. Syst.* **2017**, *29*, 2063–2079. [CrossRef]
28. Moen, E.; Bannon, D.; Kudo, T.; Graf, W.; Covert, M.; Van Valen, D. Deep learning for cellular image analysis. *Nat. Methods* **2019**, *16*, 1233–1246. [CrossRef] [PubMed]
29. Nitta, N.; Sugimura, T.; Isozaki, A.; Mikami, H.; Hiraki, K.; Sakuma, S.; Iino, T.; Arai, F.; Endo, T.; Fujiwaki, Y.; et al. Intelligent Image-Activated Cell Sorting. *Cell* **2018**, *175*, 266–276.e13. [CrossRef]

30. Solli, D.R.; Jalali, B. Analog optical computing. *Nat. Photonics* **2015**, *9*, 704–706. [CrossRef]
31. Patera, G.; Horoshko, D.B.; Kolobov, M.I. Space-time duality and quantum temporal imaging. *Phys. Rev. A* **2018**, *98*, 1–9. [CrossRef]
32. Qin, X.; Liu, S.; Chen, Y.; Cao, T.; Huang, L.; Guo, Z.; Hu, K.; Yan, J.; Peng, J. Time-lens perspective on fiber chirped pulse amplification systems. *Opt. Express* **2018**, *26*, 19950. [CrossRef]
33. AILemon. Available online: https://blog.ailemon.me/2018/11/21/free-open-source-chinese-speech-datasets/ (accessed on 2 September 2020).
34. Kingma, D.P.; Ba, J.L. Adam: A method for stochastic optimization. *arXiv* **2015**, arXiv:1412.6980.
35. Guo, B.; Lei, C.; Ito, T.; Jiang, Y.; Ozeki, Y.; Goda, K. High-throughput accurate single-cell screening of Euglena gracilis with fluorescence-assisted optofluidic time-stretch microscopy. *PLoS ONE* **2016**, *11*. [CrossRef] [PubMed]
36. Lei, C.; Ito, T.; Ugawa, M.; Nozawa, T.; Iwata, O.; Maki, M.; Okada, G.; Kobayashi, H.; Sun, X.; Tiamsak, P.; et al. High-throughput label-free image cytometry and image-based classification of live Euglena gracilis. *Biomed. Opt. Express* **2016**, *7*, 2703. [CrossRef]

MDPI

St. Alban-Anlage 66

4052 Basel

Switzerland

Tel. +41 61 683 77 34

Fax +41 61 302 89 18

www.mdpi.com

Photonics Editorial Office

E-mail: photonics@mdpi.com

www.mdpi.com/journal/photonics

9 783036 541365